SAMPLING AND ANALYSIS OF ENVIRONMENTAL CHEMICAL POLLUTANTS

A Complete Guide

SAMPLING AND ANALYSIS OF ENVIRONMENTAL CHEMICAL POLLUTANTS

A Complete Guide

by

EMMA P. POPEK

Walnut Creek, California, USA

ACADEMIC PRESS
An imprint of Elsevier Science

Amsterdam Boston Heidelberg London New York Oxford
Paris San Diego San Francisco Singapore Sydney Tokyo

Academic Press
An imprint of Elsevier Science
525 B Street, Suite 1900
San Diego, CA 92101-4495, USA

First edition 2003
ISBN: 0-12-561540-X

Library of Congress Cataloging-in-Publication Data
Popek, Emma P.
    Sampling and analysis of environmental chemical pollutants: a complete guide / by
Emma P. Popek.– 1st ed.
        p. cm
    Includes bibliographical references and index.
    ISBN 0-12-561540-X (alk. paper)
        1. Pollutants–Analysis. 2. Environmental sampling. I. Title.
    TD193.P665 2003
    628.5'2'0287–dc21

                                2003049098

British Library Cataloguing in Publication Data
Popek, Emma P.
Sampling and analysis of environmental chemical pollutants: a complete guide
1. Environmental sampling 2. Environmental sampling - Methodology
3. Pollutants - Analysis   I. Title
628.5
ISBN 012561540X

CONTENTS

INTRODUCTION

An environmental project is a complex, multi-step, and multidisciplinary under-taking. It engages many participants, who, while having a common goal, may have different or even conflicting priorities or agendas. For example, the priority of the property owner is to have the most cost-effective and efficient site investigation or remediation. The consulting company's first priority is to meet the client's demands for 'faster, cheaper, better' investigation or cleanup and at the same time to be profitable. The members of the regulatory community, who oversee the project, are usually not concerned with the cost of the project execution, as their priority is to enforce existing environmental regulations.

Various professionals who participate in the project execution also see it from different angles. Environmental projects require technical expertise in different disciplines, such as geology, chemistry, engineering, environmental compliance, and risk assessment. Each professional makes a partial contribution to the project overall effort and often has a narrowly focused view of the project, which is determined by the technical specialty of the participant and his or her role in the project organization. A geologist, who collects samples at the site, may have a poor understanding of analytical laboratory needs and capabilities, and the laboratory, in turn, has no knowledge of the sampling strategy and conditions at the project site or of the purpose of sampling. When participants focus solely on their respective tasks without taking a broader view of the project objectives, the projects tend to lose direction and as a result may arrive at erroneous conclusions. When the project objectives have not been clearly defined and understood by all members of the project team in the early stages of project execution, the project has an increased risk of failure. That is why it is essential for all team members to have a common understanding of the project's broader goals as well as a sense of what the other participants in the projects are trying to accomplish.

An understanding of the project objectives comes from initial systematic planning and continuous sharing of information during project execution. Many projects lack precisely these features, and consequently limp along in a manner that is confusing and frustrating for all participants. By analogy, a properly planned and well-executed project can be compared to a ball that moves in a straight line from point to point. By contrast, a poorly designed and badly coordinated project is akin to a multifaceted polygon, which has a general ability to roll, but its rolling swerves, lingers, or stops altogether. Needless to say, this is not an efficient way to do business.

To improve their skills in project planning and implementation, the team members seek guidance from recognized subject matter experts, such as the United States Environmental Protection Agency (EPA), Department of Defense (DOD), and Department of Energy (DOE). However, at this point they encounter an impenetrable thicket of published material, the relevance of which is not immediately clear. The enormous volume of this literature is by itself a deterrent to many professionals; the difficulty in identifying relevant information and interpreting it in practical terms is another obstacle that curbs its usefulness. Nevertheless, within the lengthy EPA, DOD, and DOE documents lies a body of core material, to which we refer daily. It takes many years of experience to orient oneself and to correctly interpret the legalistic, vague, and sometimes contradictory language of the environmental field's technical literature.

The premise of this Guide is to digest this core material into a single reference source that addresses the basic aspects of environmental chemical data collection process, such as the systematic planning, sensible field procedures, solid analytical chemistry, and the evaluation of data quality in context of their intended use. This core material, although based on the USA standards, is of a universal nature, because wherever environmental chemical data are being collected throughout the world, they are collected according to common principles. The Guide emphasizes the data collection fundamentals that are applicable to every environmental project. On a practical side, the Guide offers field procedures for basic sampling techniques that can be used in daily work. These procedures have been adapted from the EPA, DOD, and DOE protocols and revised to reflect the practical field conditions, the current state-of-the-art technology, and industry standards. The Guide will enable practicing professionals to gain a wider view of the complexities of project work and to better understand the needs and priorities of other project participants. To students of environmental data collection, the Guide offers a comprehensive view of project work, step-by-step detailed procedures for common field sampling tasks, and a wealth of practical tips for all project tasks.

ACKNOWLEDGMENTS

My foremost appreciation goes to my husband John whose selfless support and understanding made this project possible and to my son Lev for his encouragement and editorial comments. I wish to thank my mother for her interest in the progress of this effort. A special credit is due to my reviewers Dr. Garabet Kassakhian and Rose Condit for reinforcing my resolve and for technical advice and critique. I wish to acknowledge Dr. David Ben-Hur and Dr. Randy Jordan for helping in the proposal process and all my friends and family members, who cheered for me in this venture.

1

The sample and the error

1.1 THE CONCEPT OF RELEVANT AND VALID DATA

One of the main concerns of any environmental project is the collection of *relevant and valid data*. These are the data of the type, quantity, and quality that are appropriate and sufficient for the project decisions. The standards for data relevancy and validity stem from the intended use of the data since different uses require different type, quantity, and quality of data. For example, the data requirements for a risk assessment project are drastically different from those of a waste disposal project; the requirements for site investigation data are different from these for site closure.

Data relevancy and validity are two different concepts, a fact that is not always recognized by all project participants. Data can be perfectly valid, and yet irrelevant for their intended use. Conversely, the quality of data may be flawed in some way, as is usually the case in the real world, nevertheless they can be used for project decisions.

How do the project participants assure that the collected data are of the correct type, quantity, and quality, or in other words are relevant and valid? The EPA gives us general guidance in achieving this goal. To address different aspects of the data collection process, the EPA divides it into three distinctive phases: *planning*, *implementation*, and *assessment* (EPA, 2000a). Within each phase lie the data collection tasks designed to assure the relevancy and validity of the collected data. If all of the tasks within each of the three phases are conducted properly, the collected data will be not only valid and relevant to the project objectives, but they will also be indestructible, like the most durable structures on Earth, the pyramids. That is why the data collection process is best depicted as a pyramid, constructed of three ascending layers, representing the three phases of the data collection process, planning, implementation and assessment, as shown in Figure 1.1.

Each of the three phases contains several major tasks that are conducted consecutively. The data collection pyramid is built from the bottom up through carrying out these seven tasks. Take one pyramid layer or one task out, and the whole structure becomes incomplete.

1.1.1 Planning

The foundation for the collection of relevant and valid data is laid out in the planning phase through the completion of Task 1—Data Quality Objectives (DQOs) Development and Task 2—Sampling and Analysis Plan (SAP) Preparation. The DQO process enables the project participants to come to the understanding of the

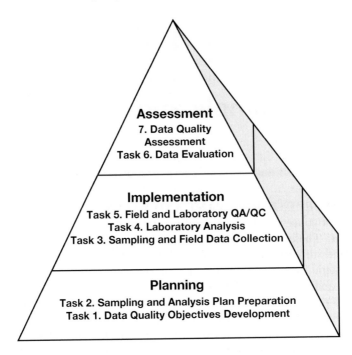

Figure 1.1   Three phases of environmental chemical data collection.

intended data use and to identify the required type, quantity, and quality of data for each separate use. In other words, *the requirements for the collection of relevant and valid data are defined through the DQO process in the planning phase of the data collection pyramid.* The project DQOs; the sampling design; and the type, quantity, and quality of data are stated in the SAP together with the requirements for field and laboratory quality control (QC), quality assurance (QA), and data evaluation procedures necessary to achieve these objectives. The SAP requirements define the course of the implementation and assessment phases of the data collection process.

1.1.2   Implementation
The implementation phase of the data collection process takes place in the field and at the laboratory where the SAP provisions are implemented. This phase consists of Task 3—Sampling and Field Data Collection, Task 4—Laboratory Analysis, and Task 5—Field and Laboratory QA/QC. In theory, if all sampling, analysis, and QA/QC tasks of this phase are conducted to the requirements of the SAP, the collected data will be relevant and valid. In reality, most projects have variances from the requirements specified in the SAP, hence a need for the next phase of data collection, i.e. assessment.

1.1.3   Assessment
The assessment phase offers us the tools to verify that the data are truly relevant and valid. The assessment phase includes Task 6—Data Evaluation and Task 7—Data

Quality Assessment. Data validity is established through the application of data evaluation procedures; their relevancy for making project decisions is determined in the course of the data quality assessment (DQA) process.

Like building the pyramids, the three-phase collection process of relevant and valid data is a daunting undertaking that can be best achieved by placing one stone at a time through a stepwise approach to each task.

1.2   IMPORTANCE OF A STEPWISE APPROACH

The EPA acknowledged the value of a stepwise approach to the planning phase when it proposed the DQO process as a tool for determining the type, quality, and quantity of data that would be sufficient for valid decision-making (EPA, 1986; EPA, 2000a). The EPA visualized and defined the DQO process as a seven-step system, with each step addressing a specific facet of the future data collection design. The DQO process enables us to define the intended use of the data; the data quality appropriate for the intended use; the sampling rationale and strategy; and many other important aspects of data collection.

The seven steps of the DQO process mold our reasoning into a logical and informed approach, turning project planning into a structured and consistent process. The Guide's Chapter 2 describes the seven steps of the DQO process and, using examples, explains how to apply them for planning of environmental data collection. Once understood, the DQO process can be used not only for the planning of environmental projects, but in any planning that we may face in everyday life. It is nothing more than a common sense planning instrument that promotes the gathering of information needed for educated decision-making; the logical thinking in decision-making; and the evaluation of risks associated with every decision.

The concept of a seven-step process is the underlying theme of this Guide. Any task, insurmountable as it may seem at a first glance, becomes feasible once broken into small, consecutive, and achievable steps. Each step has its own objectives, which we can meet without difficulty if we rely on clear and well-presented instructions to guide us through each detail of the step. To facilitate the collection of environmental chemical data, the Guide divides the data collection phases into distinctive tasks as shown in Figure 1.1; dissects the tasks into manageable steps; and provides instructions and practical advice on their execution.

1.3   SEVEN STEPS OF THE SAMPLE'S LIFE

Environmental projects revolve around environmental data collection, analytical chemistry data for toxic pollutants in particular. Chemical data enable us to conclude, whether hazardous conditions exist at a site and whether such conditions create a risk to human health and the environment. We gather environmental chemical data by collecting samples of soil, water, and other environmental media at the right time and at the right place and by analyzing them for chemical pollutants. In other words, in the core of every environmental project lies an environmental sample.

Like a living organism, an environmental sample has a finite life span and is as sensitive as living matter. It contains chemicals that undergo changes when exposed to light, air, or to variations in temperature and pressure; it houses live bacterial colonies, and contains bubbling gases or radioactive nuclei. Physical and chemical processes constantly take place in the sample, and in this chapter we shall respectfully call it 'the Sample,' for it lives and dies like a living organism.

The life of the Sample consists of seven consecutive periods or steps illustrated by a diagram in Figure 1.2. These steps are irreversible, and mistakes made by the Sample's handlers at any of them may prove fatal for the Sample's life.

Once conceived during the planning phase of the project (Step 1), the Sample gestates within the environment for an indefinite period of time until we identify the sampling point in the field (Step 2) and collect the Sample (Step 3). Identifying a correct sampling point is a critical issue, as a sample collected from a wrong location may cause the whole project to go astray. The Sample is born once it has been collected and placed in a container (Step 3). It is then transferred to the laboratory (Step 4). Mistakes made by any of the Sample's handlers at Steps 3 and 4 have a potential to invalidate the Sample and cause its premature demise. An improper sampling technique or a wrong sample container can make the Sample unsuitable for

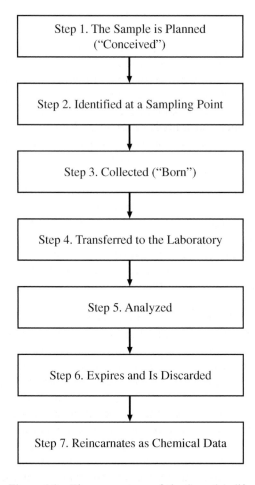

Figure 1.2 The seven steps of the Sample's life.

analysis. In fact, incorrect sample containers, packaging, and shipping during transfer from the field to the laboratory are the leading cause of sample invalidation.

The Sample is analyzed at the laboratory (Step 5), and in the course of analysis it often ceases to exist (Step 6), being completely used up in the analytical procedure. Step 5 conceals many dangers for the Sample's wellbeing, for laboratory mistakes in preparation or analysis may threaten the production of the desired data and render the Sample useless. If not completely used up during analysis, the Sample nevertheless expires (Step 6) when it reaches the limit of its holding time as prescribed by the analytical method. The Sample thus reaches the end of its life and is discarded. However, its spirit is reincarnated in the form of chemical data (Step 7). In the afterlife of the Sample, chemical data becomes the Sample's *alter ego* and acquire a life of their own. Whether they are valid or invalid, the chemical data are the Sample's immortal legacy and a testimony to our ability to plan and execute environmental projects.

1.4 TOTAL ERROR AND ITS SOURCES

As we already know, an environmental sample is a fragile living matter that can be severely damaged at every step of its existence. Due to the inherent nature of environmental media and a host of potential errors associated with sampling, analysis, and data management, the collection of environmental chemical data is not an exact science. In fact, all environmental chemical data are only the estimates of the true condition that these data represent. In order to make these estimates more accurate, we must examine the sources of errors and take measures to control them.

Total error is a combination of various errors that may occur during sampling, analysis, and in data management. It is important to understand that total error cannot be completely eliminated; it can only be controlled and minimized. Total error cannot be expressed in a convenient numeric manner. It is a concept that comprises quantifiable, statistical parameters and factors that can be described only in qualitative terms. The components of total error can be grouped into two general categories, **sampling** and **non-sampling errors**. Some sampling errors originate in the planning phase of the data collection process, whereas others are made during the implementation phase. Non-sampling errors may take place in the implementation and assessment phases.

Certain types of errors that may take place during data collection are obvious and manageable; however, the most damaging errors are hidden from plain view. For example, we usually closely scrutinize laboratory analysis measurement errors because they are easy to understand, quantify, and control. At the same time, we need to recognize the fact that errors made during sampling may be far more consequential for the data relevancy and validity than laboratory measurement errors. Such errors originate from poor judgment, negligence, lack of knowledge or simple mistakes, and because they cannot be expressed mathematically, they are described in qualitative terms. These qualitatively described errors constitute a major proportion of the total error of sampling and analysis and they easily overpower the effect of quantifiable measurement errors.

Take, for example, errors originating in the planning phase. They cannot be quantified, but their effects on data relevancy and validity could be devastating. As a result of poor planning, the purpose of data collection is often not clearly understood, and wrong types of data are collected. Incomplete or erroneous planning always negatively affects the project field implementation and produces data of inadequate quality or insufficient quantity that cannot be used for project decisions. Another source of errors that cannot be quantified is the haphazard, unfocused or poorly conducted field and laboratory implementation. It may lead to data gaps or to the production of unnecessary or unusable data. The cumulative effect of various unquantifiable errors made in planning and implementation phases erodes data validity and reduces their relevancy.

Of course, some errors are more critical than others. There are errors that are considered fatal, and errors, which while undermining the data quality to some extent, do not completely invalidate a data set. Fatal errors are errors of judgment, such as the selection of the unrepresentative sampling points in the planning process or the unjustified deviations from planned field sampling designs. Errors of negligence or oversight (a wrong analysis applied to a sample or a wrong sample selected for analysis; errors in calculations during data reduction; reporting errors; missing data) can also be fatal.

1.4.1 Sampling error

The sources of *sampling error*, which is a combination of quantifiable components and qualitative factors, can be divided into four groups:

- Errors originating from inherent sample variability
- Errors originating from population variability
- Sampling design errors
- Field procedure errors

The quantifiable components of sampling error are the natural variability within a sample itself and the variability between sample populations that may be randomly selected from the same sampled area. Both of these errors can be evaluated quantitatively in the form of statistical variance and controlled to some extent through the application of appropriate sampling designs.

The qualitative components of sampling error are the following errors in the planning process and in the implementation of field procedures:

- *Sampling design error*—the collection of excessive or insufficient amount of data; the unrepresentative selection of sampling points; wrong time of sampling that may affect data relevancy; the choice of improper analysis
- *Field procedure error*—misidentified or missed sampling points in the field; the failure to use consistent sampling procedures; gaps in field documentation; the use of incorrect sampling equipment and sample containers; incorrect sample preservation techniques or storage

The key factor in minimizing the overall sampling error is the elimination or reduction of these qualitative human errors in sampling design and field procedures.

They have a much more powerful effect on data relevancy and validity than errors originating from sample or population variability. In fact, a thoughtfully conducted planning process takes into account the population variability at the project site, and errors originating from sample variability are controlled by the proper implementation of the field sampling procedures.

1.4.2 Non-sampling error

Non-sampling errors can be categorized into *laboratory error* and *data management error*, with laboratory error further subdivided into *measurement, data interpretation, sample management, laboratory procedure* and *methodology errors.*

We can easily quantify **measurement error** due to existence of a well-developed approach to analytical methods and laboratory QC protocols. Statistically expressed accuracy and precision of an analytical method are the primary indicators of measurement error. However, no matter how accurate and precise the analysis may be, qualitative factors, such as errors in data interpretation, sample management, and analytical methodology, will increase the overall analytical error or even render results unusable. These qualitative laboratory errors that are usually made due to negligence or lack of information may arise from any of the following actions:

- *Data interpretation error*—incorrect analytical data interpretation producing false positive or false negative results
- *Sample management error*—improper sample storage; analysis of samples that have reached the end of their holding time; misplaced or mislabeled samples; sample cross-contamination
- *Laboratory procedure error*—the failure to use the proper preparation or analysis procedure; gaps in laboratory documentation
- *Methodology error*—the failure to use proper methods of sample preparation and analysis; the use of non-standard, unproven analytical methods

Data management errors may occur at the laboratory and at the consulting engineering office. They include incorrect computer program algorithms; transcription and calculation errors; bad field and laboratory record keeping practices resulting in data gaps; and other mistakes made due to a general lack of attention to the task at hand.

1.5 TOTAL ERROR AND DATA USABILITY

As we can see, total error consists for the most part of qualitative errors; the quantifiable components constitute only its minor proportion. Qualitative errors have a much greater potential to destroy the whole sampling effort than quantifiable errors, and they are usually the leading cause of resampling and reanalysis.

Clearly, every project team should be concerned with the effect of total error on data relevancy and validity and should make every effort to minimize it. Nevertheless, the cumulative effect of various sampling and non-sampling errors may erode the data validity or relevancy to a point that the data set becomes unusable for project decisions. To evaluate the magnitude of this eroding effect on the collected data, in the third and final phase of data collection process we critically assess the data in order to answer these four questions:

1. Do the data have a well-defined need, use, and purpose?
2. Are the data representative of the sampled matrix?
3. Do the data conform to applicable standards and specifications?
4. Are the data technically and legally defensible?

Total error can undoubtedly affect the outcome of all four questions. Only after we have established that the collected data are appropriate for the intended use, are representative of the sampled matrix, conform to appropriate standards, and are technically and legally defensible, the data can be described as ***data of known quality***. These are the data can be used for project decisions.

We find the answers to the four questions in the course of the data quality assessment, which is 'the scientific and statistical evaluation of data to determine if data obtained from environmental data operations are of the right type, quality, and quantity to support their intended use' (EPA, 1997a). Part of DQA is data evaluation that enables us to find out whether the collected data are valid. Another part of the DQA, the reconciliation of the collected data with the DQOs, allows us to establish data relevancy. Thus, the application of the entire DQA process to collected data enables us to determine the effect of total error on data usability.

1.5.1   Data quality indicators and acceptance criteria

When discussing data quality as a concept, we need to understand that it has a meaning only in relation to the intended use of the data and when defined by a standard that is appropriate for their intended use. Data can be judged adequate or inadequate only in the context of their intended use and only within the limits of an appropriate standard. For example, a set of data, which is of inadequate quality to be used for site characterization, may be perfectly adequate for another purpose, such as waste disposal profiling. This is possible because two different standards exist for assessing the adequacy of site characterization data and waste profiling data.

How do we access the qualitative and quantitative components of total error for data sets that consist of analytical results and an assortment of field and laboratory records? And how do we define a standard appropriate for the intended data use?

To encompass the seemingly incompatible qualitative and quantitative components of total error, we evaluate them under the umbrella of so-called data quality indicators (DQIs). DQIs are a group of quantitative and qualitative descriptors, namely ***precision, accuracy, representativeness, comparability***, and ***completeness***, summarily referred to as the PARCC parameters, used in interpreting the degree of acceptability or usability of data (EPA, 1997a). As descriptors of the overall environmental measurement system, which includes field and laboratory measurements and processes, the PARCC parameters enable us to determine the validity of the collected data.

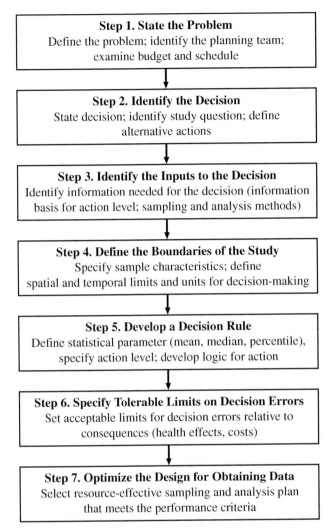

Figure 2.2   The seven steps of the Data Quality Objectives process (EPA, 2000a).

projects. Probabilistic sampling designs are typically used for site assessment, waste management, remedial investigation/feasibility studies, remedial action projects, and risk assessment studies. Data collection designs for non-probabilistic (judgmental and regulation-prescribed sampling) are not statistically based, and the DQO process does not apply to them in its entirety. Such projects may not require to be defined through a complete DQO process, nevertheless, the application of the DQO process even on a reduced scale will facilitate logical thinking and provide a systematic and organized approach to project planning.

The DQO process sounds very complicated and even intimidating, however when explained in plain terms, it becomes understandable and clear. Basically, it is a process that consists of asking questions and finding the answers. It can be simply described by using Rudyard Kipling's parable from *Just So Stories*:

'I keep six honest serving-men
(They taught me all I knew);
Their names are *What* and *Why* and *When*
And *How* and *Where* and *Who*.

I send them over land and sea,
I send them east and west;
But after they have worked for me,
I give them all a rest.'

In the following sections we will first discuss the overall goals of each step of the DQO process using the EPA's guidance and terminology (EPA, 2000a) and then review two case studies of different scopes and objectives to illustrate each step's activities and outputs. The six honest serving-men will help us in understanding this process.

Case Study A	Case Study B
Remedial Investigation at a Herbicide Storage and Packaging Facility— Probabilistic Design	**Compliance Monitoring—Regulation- Prescribed Sampling**
As shown by a previous study, surface soil at a former herbicide packaging facility has been contaminated with polychlorinated dibenzodioxins (PCDDs) and polychlorinated dibenzofurans (PCDFs), which are byproducts of chlorinated herbicide production. These chemicals are known to pose a threat to human health and the environment. The facility has been dismantled and the site is not being used for human activities, however, it is located in an ecologically sensitive area. The project scope consists of delineating the areas with contaminated soil, removing such soil, and backfilling the excavated areas with clean soil. Due to a high disposal cost, the volume of dioxin-contaminated soil is to be minimized. The work will be conducted under direct oversight from the EPA.	Due to past practices at a solvent manufacturing facility, groundwater at the site has been contaminated with chlorinated solvents. An existing groundwater treatment system removes volatile organic compounds (VOCs) with a granular activated carbon (GAC) vessel train. The treated water is discharged into a nearby creek. The local Water Quality Board has issued a National Pollutant Discharge Elimination System (NPDES) permit for the treatment system. The project scope consists of the operation and maintenance of this treatment system; the project will be conducted under a general oversight of the Water Quality Board.

2.1.1 Step 1—State the problem

The first step of the DQO process gives us direction for assembling an effective planning team, for defining the problem, and for evaluating the necessary resources. This is achieved through the following actions:

1. Identify the planning team members, including decision-makers
2. Describe the problem
3. Develop a conceptual model of the environmental hazard to be investigated
4. Determine resources, such as budget, personnel, and schedule

The DQO planning team for a large project would typically consist of a project manager and representatives of the client, regulatory agencies, technical staff, and other stakeholders, such as citizens groups or non-profit environmental organizations. The team clearly defines the environmental problem in regulatory and legal context (and sometimes in a context of local politics); identifies the source and amount of funding and available resources; and sets deadlines for various phases of project execution. The team reviews and summarizes any existing site data or a site condition that had a potential to create a threat to human health or the environment. This step often includes record searches; the examination of past and existing practices at the site; and interviews with individuals familiar with the site history and background. As a reference, the team may also evaluate the appropriateness and usefulness of the sampling and analysis methods previously used at the site or at sites with similar conditions. At this step, a complex problem may be logically broken into a set of smaller, more manageable ones, and the team will develop separate DQOs for each of them.

The outputs of Step 1 illustrated for the case studies in a box on the next page include a *list of the planning team members and their roles; identification of decision makers; a concise description and a conceptual model of the environmental problem in question; and a summary of available resources and relevant deadlines for the project, such as the budget, personnel, and schedule.*

2.1.2 Step 2—Identify the decision

In Step 2 of the DQO process the planning team will develop a decision statement that describes the purpose of the project. To formulate a decision statement, the project team will perform the following activities:

1. Identify the principal question that the project will resolve
2. Define alternative decisions
3. Develop a decision statement
4. Organize multiple decisions

The team formulates the principal question, which the planned data collection is expected to resolve, and the alternative actions that may be taken upon the resolution

Step 1. State the Problem	
Case Study A	**Case Study B**
1. Who are the members of the planning team?	
The landowner, the EPA representative, the project manager, the statistician, and the chemist, experienced in sampling, analysis, and field and laboratory QA/QC.	The facility manager, the project manager, and the chemist experienced in sampling, analysis, and field and laboratory QA/QC. The stakeholders are a group of local citizens, who are concerned with the quality of their drinking water.
2. Who is the primary decision-maker?	
The EPA representative	The facility manager
3. How is the problem described and what is a conceptual model of the potential hazard?	
• The problem is to identify and remove the soil that poses a threat to sensitive ecological receptors at the site. • The conceptual model describes a former herbicide packaging facility, where chlorinated herbicides were stored and repackaged outdoors 50 years ago. Due to careless practices, these herbicides in a powder form were spread over a large surface area. Polychlorinated dioxins and furans, the recalcitrant byproducts of herbicide production, have been identified in the surface soil at the site. They may present a risk to the habitat of sensitive ecological species in the near vicinity.	• The problem is to determine whether the treated water meets the limitations of the NPDES permit. • The conceptual model describes leaking storage tanks as the source of groundwater contamination at a chlorinated solvent manufacturing facility. Contaminant concentrations in groundwater exceed the regulatory standards. The groundwater plume has been characterized and mapped. The plume threatens the drinking water wells located down gradient from the plume. Two years ago a treatment system was built at the source of the plume, which has been removing groundwater from the subsurface, pumping it through the GAC vessel train, and discharging it into a creek.
4. What are the available resources and when will the work be conducted?	
A contractor company represented in the DQO process by the project manager, the chemist, and the statistician will conduct project planning. The contractor company will provide the necessary resources and perform the work. The project budget and schedule are negotiable.	A consulting company represented in the DQO process by the project manager and the chemist will perform the work; the consultant will provide personnel for the operation of the treatment system. The project has a limited budget and will last for two years.

of the principal question. The team then combines the principal question and the alternative action into a decision statement, which summarizes the purpose of the project. Multiple decision statements may be needed for various project phases, and the planning team will establish their logical order and relationship.

The Step 2 output is *an established purpose of the project in a decision statement linking the principal question to the possible actions that will solve the problem.*

Step 2. Identify the Decisions	
Case Study A	**Case Study B**
1. What is the principal question?	
Do the PCDD/PCDF concentrations in soil pose a risk to the environment?	Do contaminant concentrations in treated water exceed the NPDES permit limitations?
2. What are alternative actions that could result from resolution of the principal question?	
1. If the PCDD/PCDF concentrations in soil pose a risk to the environment, remove the ecological receptors from the site—not a feasible action. 2. If the PCDD/PCDF concentrations in soil pose a risk to the environment, remove the soil that is found to be hazardous to ecological receptors. 3. If the PCDD/PCDF concentrations in soil *do not* pose a risk to the environment, leave the soil at the site.	1. If the NPDES permit limitations are exceeded, the operation of the treatment system continues—not a feasible action. 2. If the NPDES permit limitations are exceeded, the operation of the treatment system is stopped and the GAC vessels are replaced. 3. If the NPDES permit limitations are *not* exceeded, the operation of the treatment system continues.
3. Combine the principal study question and the alternative action into a decision statement.	
Soil with PCDD/PCDF concentrations that may present a risk to ecological receptors will be removed from the site; soil with concentrations that do not present such risk will remain at the site.	The operation of the treatment system will proceed uninterrupted until the treated water exceeds the NPDES permit limitations, upon which the operation will stop and the GAC vessels will be replaced.
4. Organize multiple decisions.	
Only one decision is being evaluated.	Only one decision is being evaluated.

2.1.3   Step 3—Identify inputs to decisions

In Step 3, the planning team will identify the types of information required for answering the principal question defined in Step 2 and determine the availability of appropriate data collection methods. The team will conduct the following activities:

1. Identify the information needed
2. Determine sources for this information

3. Determine the basis for the selection of the action level
4. Identify the necessary sampling and analysis methods

The planning team defines the types of information required to resolve the principal question, such as chemical, toxicological, biological, geological data, and their sources. The sources of information may also include previously collected data; historical records; regulatory guidance; risk assessment conclusions; consensus opinion; and published literature. The team will also decide whether the existing data are suitable and sufficient for decision-making or whether new data will be collected.

At this point, we arrive at the definition of a concept that is very important for environmental chemical data collection projects. It is the concept of the ***action level***, which is defined by the EPA (EPA, 2000a), as follows: ***The action level (C_a) is the threshold value, which provides the criterion for choosing between alternative actions***.

Depending on the nature of the project, different conditions, such as current or future land use, the presence of sensitive ecological and human receptors, or applicable environmental regulations, will affect the action level selection. The action level is a contaminant concentration, for example, a regulatory standard, a risk assessment result, a background or ambient concentration, a value that is based on a detection limit of an analytical method, or a negotiated, site-specific value. In Step 3, the planning team identifies the possible action level and references its source. If the action level is based on a regulatory requirement, the planning team will immediately know its numeric value. However, a discharge permit or a risk-based action level might not be yet available to the planning team at this time.

The team identifies the contaminants of concern and proposes potentially suitable measurement methods. The measurement methods may come from the EPA guidance manuals for analytical methods or from other appropriate sources for standard methods of analysis. Depending on the type of data, Performance Based Measurement Systems (PBMS) may also be used for obtaining physical, chemical, or biological measurements. The PBMS are alternative analytical methods that are developed for a specific analyte or a group of analytes and offer an innovative or a more efficient analytical approach being at the same time more cost-effective than the standard methods of analysis.

The analytical method selection is an intricate part of the DQO process because of the variety of existing analytical methods and techniques. A chemist experienced in environmental analysis should make the selection using the action level as a starting point and refining the choice based on other aspects of the project DQOs.

The outputs of Step 3 include *a list of contaminants of concern to be measured; a list of information sources and methods that indicate how each action level will be derived; a list of information that provides continuity with the past and future work, such as databases, survey coordinates, uniform sample numbering systems; and identification of the appropriate analytical methods to meet the action level requirements.*

Step 3. Identify Inputs to Decisions	
Case Study A	**Case Study B**
1. What kind of information will be required to resolve the decision statement?	
PCDD/PCDF in soil are the contaminants of concern for this project. The required information is the concentrations of PCDD/PCDF in soil, expressed in the units of 2,3,7,8-tetrachlorodibenzodioxin (2,3,7,8-TCDD) toxicity equivalents (TEQ). The TEQ is a calculated value, which contains all of the PCDD/PCDF homologue concentrations factored in according to their toxicity.	VOCs are the contaminants of concern for this project. The required information is the VOC concentrations in the effluent water stream. To support decisions related to the treatment system operation, VOC concentrations in influent samples are also required.
2. What are the sources for each item of information identified?	
New chemical data will be collected to determine the PCDD/PCDF concentrations in soil. The coordinates of each sampling location will be surveyed for future entry into a database together with chemical data.	To verify discharge permit compliance, VOC concentrations will be determined in the treatment system effluent. Influent samples will be also analyzed to calculate the GAC vessel train loading. An existing sample numbering system will be used and the data will be entered into an existing database.
3. What is the basis for the action level?	
The action level for contaminated soil removal will be derived from an ecological risk assessment.	The NPDES permit limitations will constitute the action levels for this project.
4. What sampling and analysis methods are appropriate?	
The appropriate method for PCDD/PCDF analysis is the EPA Method 8290 of the Test Methods for Evaluating Solid Waste, SW-846 (EPA, 1996a).	According to the NPDES permit, effluent and influent water samples will be analyzed for VOCs by EPA Method 624 (EPA, 1983).

2.1.4   Step 4—Define the study boundaries

In Step 4, the planning team will define the geographic and temporal boundaries of the problem; examine the practical constraints for collecting data; and define the sub-population for decision-making. To achieve this, the team will perform the following activities:

1. Define target population of interest
2. Specify the spatial boundaries for data collection
3. Determine the time frame for collecting data and making the decision

4. Determine the practical constraints on collecting data
5. Determine the smallest sub-population area, volume or time for which separate decisions will be made

The target population is a set of all samples for which the decision-maker wants to draw conclusions. It represents environmental conditions within certain spatial or temporal boundaries. For environmental projects, the target populations are usually samples of surface and subsurface soil, groundwater, surface water, or air, collected from a certain space at a certain time.

In Step 4, the planning team determines the spatial boundaries of the area to be sampled. The area may be, for example, a plot of land of known size, where samples will be collected at discrete depths; a point source, such as a water faucet; or a well field from which samples will be collected at a known frequency. The team defines the time frame during which the decision applies and the time when the data will be collected. It also identifies any practical constraints that may interfere with the project execution, such as adverse seasonal conditions or natural or man-made physical obstacles. Among the practical constraints are financial limitations, which are always a consideration in developing sampling designs. Examples of the temporal boundaries or time limitations are possible changes in site conditions or short field seasons in some geographic areas.

The target population may be broken into sub-populations for which a separate decision will be made. Examples of sub-populations may be samples of soil collected from a small surface area within a large area to be sampled; groundwater samples collected from discrete depths; or soil samples collected from a section of a stockpile. By clearly defining the target population and the sub-populations, the planning team lays a foundation for the sampling design and future data interpretation. The team will establish the scale for decision-making by defining the spatial and temporal boundaries of the sub-populations. Such spatial boundaries may include definitions of the sampling grid sizes or vertical and lateral excavation boundaries.

The outputs of Step 4 are *detailed descriptions of the characteristics that define the population to be sampled; detailed descriptions of geographic limits (spatial boundaries); time frame for collecting data and making the decision (temporal boundaries); the scale for decision making (the size of the target population and the sub-population); and a list of practical constraints that may interfere with data collection.*

Step 4. Define the Study Boundaries	
Case Study A	**Case Study B**
1. What population will be sampled?	
The target population is the samples from the upper 12 inches of soil in the contaminated area.	The target population is water samples that represent the volumes of discharged treated water.

Step 4. Define the Study Boundaries (Continued)	
Case Study A	**Case Study B**
2. What are the spatial boundaries that the data must represent or where will the sampling be conducted?	
The contaminated site is a rectangle 500 yards long and 350 yards wide. Individual soil samples from 2 to 4 inches below ground surface (bgs) and from shallow subsurface (at 12 inches bgs) will be collected using a systematic grid approach. Several samples will be collected from each grid unit.	Each sample will represent the volume of water treated and discharged within the period of time between two sampling events.
3. What are the temporal boundaries that the data must represent or when will the sampling be conducted?	
The collected data will reflect the existing site conditions. The data collection activities will start after the project is mobilized and will continue for approximately 6 weeks.	The collected data will characterize the treated water over a two-year period of the treatment system operation. Samples will be collected at regular time intervals as specified in the NPDES permit.
4. What is the scale for decision-making?	
Sub-population: a decision on whether to remove soil will be made for *each individual grid unit* based on the mean TEQ concentration in every grid unit. *Target population*: a decision on whether the action level has been met will be made after soil with the mean TEQ concentration exceeding the action level has been removed from each grid unit and it will be based on a statistical evaluation of the data collected from the entire 500 yard by 350 yard area.	In this case, sub-population is the same as the target population: a decision will be made every time an effluent sample data are obtained.
5. What are the practical constraints for data collection?	
A practical constraint for data collection is the fact that a part of the site is currently used as a staging area for old equipment and is not available for sampling. There are also time constraints for the field season. Budgetary constraints require that the smallest possible volume of soil be removed due to a high cost of incineration of dioxin-contaminated soil.	There are no practical constraints for data collection, although budgetary constraints exist.

2.1.5   Step 5—Develop the decision rule

In Step 5 of the DQO process, the planning team will develop a *theoretical* decision rule that will enable the decision-maker to choose among alternative actions. The following activities are part of Step 5:

1. Specify an appropriate statistical parameter that will represent the population, such as mean, median, percentile
2. Confirm that the action level exceeds measurement detection limits
3. Develop a theoretical decision rule in the form of 'if...then' statement

The planning team specifies the *parameter of interest*, which is the statistical parameter that characterizes the population. The parameter of interest provides a reasonable estimate of the true contaminant concentration, and it may be the mean, median or percentile of a statistical population.

The team also states the action level in numerical terms and verifies that the proposed measurement methods have adequate sensitivity to meet the stated action levels. Finally, the team develops a decision rule expressed as an *'if...then'* statement that links the parameter of interest, the scale of decision-making, the action level, and the alternative action.

To explain the concept of decision rule, we will use an example of a scenario often encountered at sites that undergo environmental cleanup. We will return to this example in the following chapters to illustrate other concepts used in the DQO process.

Example 2.1:   Theoretical decision rule

We need to make a decision related to the disposition of soil that has been excavated from the subsurface at a site with lead contamination history. Excavated soil suspected of containing lead has been stockpiled. We may use this soil as backfill (i.e. place it back into the ground), if the mean lead concentration in it is below the action level of 100 milligram per kilogram (mg/kg). To decide whether the soil is acceptable as backfill, we will sample the soil and analyze it for lead. The mean concentration of lead in soil will represent the statistical population parameter.

A theoretical decision rule: If a true mean concentration (μ) of lead (parameter of interest) in the stockpile (scale of decision-making) is below the action level of 100 mg/kg, then the soil will be used as backfill.

The alternative action: If a true mean concentration (μ) of lead in the stockpile is above the action level of 100 mg/kg, then the soil will not be used as backfill and will be further characterized to determine disposal options.

The decision rule assumes that perfect information has been obtained from an unlimited number of samples and that the sample mean concentration ($\bar{x}$) is equal to the true mean concentration (μ). (The definitions of sample mean and true mean concentrations can be found in Appendix 1). The reality is that we never have perfect information and unlimited data, and that is why this decision rule is only a *theoretical* one. In fact, environmental project decisions are made on data that are obtained from a limited number of samples and that are subject to various errors discussed in Chapter 1. These data are only the *estimates* of the true mean concentration. Hence, the decisions made on such intrinsically imperfect data can occasionally be wrong. Nevertheless, a theoretical decision rule, inaccurate in statistical terms as it may be, determines the general direction and the logic of the project activities and is the most significant output of Step 5 as illustrated for the case studies in the box on page 24.

2.1.6   Step 6—Specify limits on decision errors

The sixth step of the DQO process addresses the fact that the true mean contaminant concentration μ will never be known and that it can be only approximated by the sample mean concentration $\bar{x}$ calculated from the collected data. The mean sample concentration approximates the true mean concentration *with a certain error*. And therefore, a decision based on this mean concentration will also contain an inherent error. The latter is called **decision error**.

Step 6 allows us to create a statistical approach for the evaluation of the collected data. Using a statistical test and the statistical parameters selected in Step 6, we will be able to control decision error and make decisions with a certain level of confidence. Decision error, like total error, can only be minimized, but never entirely eliminated. However, we can control this error by setting a tolerable level of risk of an incorrect decision. Conducting Step 6 enables the planning team to specify acceptable probabilities of making an error (the tolerable limits on decision errors). At this step of the DQO process, the project team will address the following issues:

1. Determine the range of contaminant of concern concentrations
2. Choose the null hypothesis
3. Examine consequences of making an incorrect decision
4. Specify a range of values where the consequences of making an incorrect decision are minor (gray region)
5. Assign probability values to points above and below the action level that reflect tolerable probability for decision errors

During Step 6 of the DQO process the planning team evaluates the severity of a decision error consequences in terms of its effects on human health, the environment, and the project budget. The accepted level of risk is then expressed as a *decision performance goal probability statement*. The team is willing to accept a certain level of risk of making an incorrect decision on condition it presents an insignificant threat to human health and the environment. Financial constraints or legal and political ramifications often influence the level of risk associated with a wrong decision that the planning team would be willing to tolerate. Occasionally these considerations may overpower the considerations of human health and the environment.

The planning team will devise measures for control or minimization of decision error through controlling total error. To do this, the planning team may need the expertise of a statistician who will develop a data collection design with a tolerable probability for the occurrence of a decision error.

Depending on the project objectives, Step 6 may be the most complex of all DQO process steps. However, for a majority of environmental projects that are driven by

Step 5. Develop the Decision Rule	
Case Study A	**Case Study B**
1. What is the statistical parameter that characterizes the target population?	
The mean TEQ concentration for all samples collected from the *entire area* will be the statistical parameter that characterizes the target population. A statistical evaluation will be conducted for the TEQ concentrations for all of the samples collected from the entire area. The mean TEQ concentration for *each grid* will be compared to the action level for a 'yes or no' decision.	This project is not based on a probabilistic sampling design and does not have statistical parameters. The parameters that characterize the population of interest are specified in the NPDES discharge permit. They are the VOC concentrations in every effluent sample collected.
2. What is the numerical value for the action level?	
A risk-based action level obtained from an ecological risk assessment is 1 microgram per kilogram (µg/kg) TEQ.	The action level (discharge permit limitation) for each of the contaminants of concern is 1 microgram per liter (µg/l).
3. What decisions do we make based on obtained data?	
• If a mean TEQ concentration in a *grid unit* for the samples collected at 2 to 4 inches bgs exceeds 1 µg/kg, then the soil in this grid will be removed to a depth of 12 inches bgs. Additional samples will be collected from the bottom of the excavated area and analyzed for PCDDs/PCDFs. • If a mean TEQ concentration of a *grid unit* for the samples collected at 2 to 4 inches bgs is below 1 µg/kg, then the soil excavation in this grid unit will not be conducted. • If a mean TEQ concentration for *all samples* collected from the *entire area* exceeds 1 µg/kg, then the removal of contaminated soil 'hot spots' in individual grids will continue. • If a mean TEQ concentration for *all samples* collected from the *entire area* is below 1 µg/kg, then the removal action will stop.	• If a concentration of any single contaminant of concern in an effluent sample exceeds 1 µg/l, then operation of the treatment system will stop and the treated water will not be discharged into the creek. • If the concentrations of all of the contaminants of concern in an effluent sample are below 1 µg/l, then the operation of the treatment system will continue uninterrupted and the treated water will be discharged into the creek.

regulatory requirements and budgetary considerations, there is no need for elaborate statistics as long as all of the stakeholders fully understand the potential consequences of each decision error.

2.1.6.1 Determine the range of contaminant of concern concentrations

To determine the magnitude of the effect of total error on contaminant of concern concentrations, we need to evaluate the expected range of these concentrations. Sources for information on the range of expected concentrations may include previously collected chemical data or historical records; professional judgment based on similar project scenarios may be also used.

Knowing the range of the contaminant of concern concentrations will enable the planning team to estimate the expected mean concentration and to evaluate sample matrix variability. The planning team compares the expected mean concentration to the action level to determine a need for further activities of Step 6. If the expected range of contaminant of concern concentrations is in orders of magnitude lower or greater than the action level, there would be no need to proceed with further Step 6 activities, as the probability of making a wrong decision in this case is near zero.

Step 6. Specify Limits on Decision Errors	
Case Study A	Case Study B
1. What is the possible range of the contaminant of concern concentrations?	
A possible range of PCDD/PCDF concentrations is unknown.	Because the treatment system has been in operation for a year, the range of VOC concentrations in the effluent is well defined, and it is 0.7 to 1.1 µg/l.

2.1.6.2 Choose the null hypothesis

In Step 5, the planning team made a *theoretical* decision rule based on an assumption that the sample mean concentration obtained from analysis is equal to the true mean concentration. Therefore, decisions made according to this rule would always be correct. In reality, however, because the sample mean concentration is only an *estimate* of the true mean concentration, the decision rule will be certainly correct only when the sample mean concentration is clearly above or below the action level. But what if the sample mean concentration is close to the action level? Obviously, given the presence of total error, the closeness of these values makes the decision making process uncertain and the probability of making a wrong decision great. Nevertheless, even in these ambiguous circumstances a decision must be made, and a method to resolve this problem is offered in Step 6 of the DQO process.

To minimize the risk of making a wrong decision, the planning team will use a concept of *baseline condition* and a statistical technique called *hypothesis testing*.

The **baseline condition** is a statement that establishes the relationship between the true mean contaminant concentration and the action level. An example of the baseline condition statement is: 'The true mean concentration exceeds the action level.' The baseline condition is presumed to be true until proven false by overwhelming evidence from sample data. The statement that establishes the opposite relationship is called the *alternative condition*. An alternative condition for the stated baseline condition is as follows: 'The true mean concentration is below the action level.'

We establish the baseline and the alternative conditions during the planning process as *statements* that guide our logic in the decision-making process. Only one of these conditions can be correct, and the true one will be identified based on a statistical evaluation of the collected data, which compares the sample mean concentration to the action level. If the baseline condition is proven false by the sample data, we will reject it and accept the alternative condition as the true one.

The choice between the baseline and alternative conditions is easy if the mean concentration significantly differs from the action level. But how can we determine, which of the two conditions is correct in a situation when a sample mean concentration approximates the action level? This can be achieved by the application of *hypothesis testing*, a statistical testing technique that enables us to choose between the baseline condition and the alternative condition. Using this technique, the team defines a baseline condition that is presumed to be true, unless proven otherwise, and calls it the *null hypothesis* (H_0). An *alternative hypothesis* (H_a) then assumes the alternative condition. These hypotheses can be expressed as the following equations:

$$H_0 : \mu \geq C_a \text{ versus } H_a : \mu < C_a$$

where μ is the true mean concentration and C_a is the action level.

What is the importance of the null and the alternative hypotheses? They enable us to link the baseline and alternative condition statements to statistical testing and to numerically expressed probabilities. The application of a statistical test to the sample data during data quality assessment will enable us to decide with a *chosen level of confidence* whether the true mean concentration is above or below the action level. If a statistical test indicates that the null hypothesis is not overwhelmingly supported by the sample data with the chosen level of confidence, we will reject it and accept the alternative hypothesis as a true one. In this manner we will make a choice between the baseline and the alternative condition.

If the sample mean concentration is close in value to the action level, the decision, however, becomes uncertain, and two types of decision errors may take place:

1. *False rejection decision error*
   Based on the sample data, we may reject the null hypothesis when in fact it is true, and consequently accept the alternative hypothesis. By failing to recognize a true state and rejecting it in favor of a false state, we will make a decision error called a *false rejection decision error*. It is also called a *false positive error*, or in statistical terms, *Type I decision error*. The measure of the size of this error or the probability is named 'alpha' (α). The probability of making a correct decision (accepting the null hypothesis when it is true) is then equal to $1-\alpha$. For environmental projects, α is usually selected in the range of 0.05–0.20.

2. *False acceptance decision error*
   Based on the sample data, we may accept the null hypotheses when in fact it is false, and make a *false acceptance decision error* or a *false negative error*. In

statistical terms, it is a *Type II decision error*. The measure of the size of this error or the probability of a false acceptance decision error is named 'beta' (β). The probability of making a correct decision (rejecting the null hypothesis when it is false) is equal to $1-\beta$. The more data are collected to characterize the population, the lower the value of β.

Example 2.2 illustrates the concepts of the baseline and the alternative condition statements, and the null and alternative hypotheses; the case studies are presented on page 29.

Example 2.2: Baseline and alternative hypothesis and the two types of decision errors

We need to make a decision related to the disposition of soil that has been excavated from the subsurface at a site with lead contamination history. Excavated soil suspected of containing lead has been stockpiled. We may use this soil as backfill, if the mean lead concentration is below the action level of 100 mg/kg. To decide if the soil is acceptable as backfill, we will sample the soil and analyze it for lead. The mean concentration of lead in soil will represent the statistical population parameter.

A theoretical decision rule: If a true mean concentration (μ) of lead in the stockpile is below the action level of 100 mg/kg, then the soil will be used as backfill.

The alternative action: If a true mean concentration (μ) of lead in the stockpile is above the action level of 100 mg/kg, then the soil will not be used as backfill and will be further characterized to determine disposal options.

A baseline condition and the null hypothesis: The *true* mean concentration (μ) of lead in soil is above the action level; H_0: $\mu \geqslant 100$ mg/kg

The alternative condition and the alternative hypothesis: The *true* mean concentration (μ) of lead in soil is below the action level; H_a: $\mu < 100$ mg/kg

The null hypothesis versus the alternative hypothesis:
H_0: $\mu \geqslant 100$ mg/kg versus H_a: $\mu < 100$ mg/kg

Two types of decision errors may take place in decision making:

1. *False rejection decision error*
   Suppose that the *true* mean concentration μ is 110 mg/kg, but the *sample* mean concentration is 90 mg/kg. In this case, the null hypothesis is true (H_0: 110 mg/kg > 100 mg/kg). However, basing our decision on the sample data, we reject it in favor of the alternative hypothesis (H_a: 90 mg/kg < 100 mg/kg) and make a false rejection decision error.

2. *False acceptance decision error*
   Now suppose that the *true* mean concentration μ is 90 mg/kg, but the *sample* mean concentration is 110 mg/kg. Then the null hypothesis is false because the true mean concentration is not greater than the action level (90 mg/kg < 100 mg/kg). But we are not able to recognize this, as the sample mean concentration supports the null hypothesis (H_0: 110 mg/kg > 100 mg/kg). Based on the sample data we accept the null hypothesis and thus make a false acceptance decision error.

Table 2.1 Decision error summary

Sample mean concentration	True mean concentration	
	Baseline condition is true $(H_0: \mu \geq C_a)$	Alternative condition is true $(H_a: \mu < C_a)$
Baseline condition is true $(H_0: \mu \geq C_a)$	Correct decision The probability of making a correct decision: $(1-\alpha)$	False acceptance decision error False negative decision error Type II decision error Probability: β Risk, error rate: $100 \times \beta$
Alternative condition is true $(H_a: \mu < C_a)$	False rejection decision error False positive decision error Type I decision error Probability: α Risk, error rate: $100 \times \alpha$	Correct decision The probability of making a correct decision: $(1-\beta)$

Table 2.1 summarizes possible conclusions of the decision-making process and common statistical terms used for describing decision errors in hypothesis testing.

Although basic statistics is outside the scope of this Guide, certain statistical concepts are important for the understanding of probabilistic sampling designs. Appendix 1 contains the definitions of statistical terms and some of the equations used in the statistical evaluation of environmental data.

2.1.6.3 Examine the consequences of making an incorrect decision

Example 2.2 describes lead-contaminated soil with the baseline condition stating that the true mean concentration of lead in soil exceeds the action level. If a false rejection decision error has been made, contaminated soil with concentrations of lead exceeding the action level will be used as backfill, and will therefore continue to pose a risk to human health and the environment. On the opposite, as a consequence of a false acceptance decision error, soil with lead concentrations below the action level will not be used as backfill and will require unnecessary disposal at an additional cost.

In this example, the two decision errors have quite different consequences: a false rejection decision error has consequences that will directly affect the environment, whereas a false acceptance decision error will cause unnecessary spending. Recognizing a decision error with more severe consequences is a pivotal point in the DQO process and a dominating factor in optimizing the data collection design.

If the consequences of a decision error are significant for human health and the environment, the planning team needs to reduce its probability by taking measures to minimize total error and its components. To do this, the planning team may implement a quality system that controls the components of total error. For example, sampling design error can be minimized through collecting a larger number of samples or by using a more effective design for more accurate and precise representation of the site population. To control measurement error, the planning team may select analytical methods different from those originally proposed in Step 3, and establish more stringent criteria for the PARCC parameters.

Step 6. Specify Limits on Decision Errors	
Case Study A	**Case Study B**
1. What decision errors can occur?	
The two possible decision errors are as follows: • Deciding that the true mean TEQ concentration in soil exceeds the action level when it does not • Deciding that the true mean TEQ concentration in soil is below the action level when it actually is not	The two possible decision errors are as follows: • Deciding that the VOC concentrations in treated water exceed the discharge limitations when they do not • Deciding that the VOC concentrations in treated water do not exceed the discharge limitations when they actually do
2. What are the baseline and the alternative conditions? What are the null and the alternative hypotheses?	
• The baseline condition and the null hypothesis are 'The true mean TEQ concentration of the entire area exceeds the action level of $1\,\mu g/kg$ or $H_0\!:\mu \geq 1\,\mu g/kg$.' • The alternative condition and the alternative hypothesis are 'The true mean TEQ soil concentration of the entire area is below the action level of $1\,\mu g/kg$ or $H_a\!:\mu < 1\,\mu g/kg$.' • The *false rejection decision* error will occur if the null hypothesis is rejected when it is true, and the alternative hypothesis is accepted. In this case, the soil will be deemed non-hazardous when, in fact, it is hazardous. • The *false acceptance decision error* will occur if the null hypothesis is accepted when it is false; consequently, the non-hazardous soil will be considered hazardous.	Because mean concentrations will not be used for decision-making, we cannot apply hypothesis testing for this project. Nevertheless, we can use hypothesis testing terms and logic to evaluate possible decision errors. • The baseline condition is 'The VOC concentrations in the effluent sample are below the discharge limitation of $1\,\mu g/l$.' • The alternative condition is 'The VOC concentrations in the effluent sample are above the discharge limitation of $1\,\mu g/l$.' • The *false rejection decision error* will occur if the baseline condition is rejected when it is true, and the alternative condition is accepted. The VOC concentrations in the effluent will be believed to exceed the discharge limitation when, in fact, they will not. • The *false acceptance decision error* will occur if the baseline condition is accepted when it is false. The VOC concentrations in the effluent will be incorrectly believed to be below $1\,\mu g/l$.

Clearly, collecting more data to obtain a more accurate estimate of the true mean concentration will reduce decision error risk but it also would increase the project costs. There is always a fine balance between the need to make a valid decision and the need to stay within an available or reasonable budget, and sometimes the two needs may be in conflict. As a compromise, after evaluating the severity of decision error consequences, the planning team may consider adopting a less comprehensive sampling design, wider acceptance criteria for analytical data quality indicators, or wider tolerable limits on decision errors.

Step 6. Specify Limits on Decision Errors	
Case Study A	**Case Study B**
1. What are the potential consequences for each decision error?	
• The consequence of *a false rejection decision error* (deciding that soil is not hazardous when, in fact, it is) will be that it remains at the site and continues to be a threat to the ecosystem. • The consequence of *a false acceptance decision error* (deciding that the soil is hazardous when, in fact, it is not) will be a financial loss. Removing more soil than necessary will increase the project disposal costs.	• The consequence of *a false rejection decision error* (deciding that the concentrations in treated water exceed the discharge limitations when, in fact, they do not) will be a financial loss. The subsequent shutdown of the treatment system operation and the replacement of GAC vessels are unnecessary expenses to the project. • The consequence of *a false acceptance decision error* (erroneously deciding that the VOC concentrations do not exceed the discharge limitations when, in fact, they do) will be an unlawful discharge to the creek and a substantial fine for the NPDES permit violation.
2. What decision has more severe consequences near the action level?	
The consequence of *a false rejection decision error* will be a more severe one because of the unmitigated threat to the environment.	The consequence of *a false acceptance decision error* will be a more severe one because of the violation of the NPDES permit.

2.1.6.4 Specify a 'gray region'

As we already know, in Step 6 of the DQO process the planning team will be mainly concerned with the uncertainty of decisions based on the mean sample concentrations that are close to the action level. Collecting a greater number of samples to better approximate the true concentration will reduce this uncertainty. However, this may not be feasible due to increased sampling and analysis costs. The uncertainty associated with decisions based on the mean sample concentrations that are close to the action level is addressed in the DQO process through a concept of the ***gray region***.

The *gray region is a range of true mean concentrations near the action level where a probability of making a false acceptance decision error is high.* In this concentration range, large decision error rates are considered tolerable because the consequences of such decision errors are relatively minor. Depending on the way the baseline condition has been formulated, the gray region may be in the range of concentrations either above or below the action level. It is bordered by the action level on one side and a *selected* concentration value on the other side. The gray region is usually expressed as a fraction of the action level, i.e. 10 percent, 20 percent or 30 percent of the action level.

The concept of gray region is best understood when presented in a graphic form. Figure 2.3 illustrates a decision performance goal diagram for Example 2.2, i.e. a decision for soil contaminated with lead, and represents a decision-making scenario for the baseline condition stating that true mean concentration of lead in soil exceeds the action level of 100 mg/kg or the null hypothesis H_0: $\mu \geq 100$ mg/kg. The gray region in this diagram is established at 20 percent of the action level and ranges from 80 to 100 mg/kg./P >

The decision performance goal diagram in Figure 2.3 reflects the fact that the difference between the *true mean* and *sample mean* concentrations may be the cause of decision errors. The dotted line represents the probability of deciding whether the true mean concentration of lead in soil exceeds the action level. For mean sample concentrations that significantly exceed the action level, the false rejection decision error rate α is extremely low, and the probability of deciding whether the true mean concentration exceeds the action level is close to 1. For example, if the sample mean concentration exceeds 140 mg/kg, the true mean concentration may most likely be assumed to be greater than 100 mg/kg. On the opposite side of the action level, with the sample mean concentrations clearly below the action level, the false acceptance error rates are low and the probability of deciding that the true mean concentration exceeds the action level is also low. For example, if the sample mean concentration is 20 mg/kg, the probability of deciding that true concentration exceeds 100 mg/kg is very low. In other words, if the sample mean concentrations are substantially greater or lower than the action level, the probability of making a false acceptance or a false

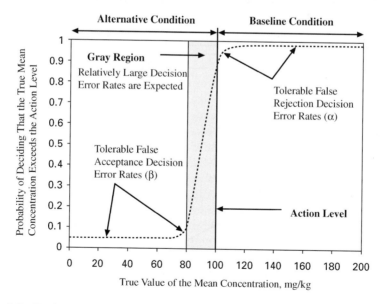

Figure 2.3 Decision performance goal diagram: true mean concentration exceeds the action level (EPA, 2000a). The dotted line represents the probability of deciding that the true mean concentration exceeds the action level.

rejection decision error is insignificant, and a decision can be made virtually without error.

However, the closer the sample mean concentrations are to the action level, the higher the probability of making a decision error. For example, if the sample mean concentration is 110 mg/kg and the true mean concentration 90 mg/kg, the probability of making a false acceptance decision error is high. A consequence of false acceptance decision error is that soil with the true mean concentration of lead below the action level will be found unsuitable for backfill and will require disposition at an additional expense to the project. Conversely, if true mean concentration of lead is 110 mg/kg, and the sample mean concentration is 90 mg/kg, a false rejection decision error is likely to be made. As a consequence, soil with the true mean lead concentration exceeding the action level will be used as backfill and continue to present a threat to human health and the environment.

Figure 2.4 illustrates the changes in the decision performance goal diagram for the opposite baseline condition (the true mean concentration of lead in soil is below the action level or the null hypothesis H_0: $\mu < 100$ mg/kg).

The probability curve is the same as in Figure 2.3, but the gray region in this case is situated on the opposite side of the action level, and the false acceptance and false rejection decision errors have changed places on the probability curve. With low probabilities of decision error, soil with sample mean concentrations significantly below 100 mg/kg will be accepted as backfill, soil with sample mean concentrations clearly exceeding 100 mg/kg will be rejected. However, if the true mean concentration is slightly above the action level, for example, 110 mg/kg, and the sample mean concentration is less than 100 mg/kg (e.g. 90 mg/kg), then the project team will be likely to make a false acceptance decision error. A consequence of false acceptance

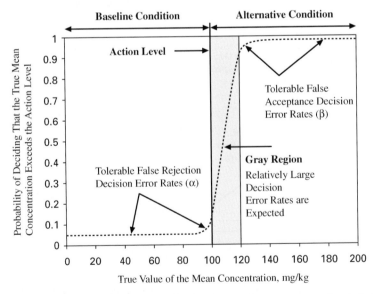

Figure 2.4 Decision performance goal diagram: true mean concentration is below the action level (EPA, 2000a). The dotted line represents the probability of deciding that the true mean concentration exceeds the action level.

decision error in this case would be quite different from the previously described one, as soil with lead concentrations exceeding the action level would be used as backfill. However, the threat to human health and the environment will be minimal because of the proximity of the sample mean concentration to the action level. Conversely, if the sample mean concentration is 110 mg/kg but the true mean concentration is 90 mg/kg, the decision makers will commit a false rejection decision error and soil suitable for backfill will be disposed of unnecessarily.

The two diagrams illustrate how the formulation of the baseline condition effects the outcome of the decision process, as quite different decisions may be made for the same data set depending on how the baseline condition has been stated. That is why the baseline condition should be re-evaluated and, if necessary, restated after the consequences of false acceptance and false rejection decision errors have been evaluated. *If the most severe decision error occurs above the action level, the baseline condition should assume that the mean concentration exceeds the action level ($H_0: \mu \geq C_a$).*

Selecting a gray region where the total error rates are high, but the consequences of making a false acceptance decision error are relatively mild in terms of the threat presented to human health and the environment is an important factor in optimizing a data collection design. Obviously, the more samples we collect and analyze, the better is the approximation of the true mean contaminant concentration; the lower the probability of making a false acceptance decision error β; and the narrower the selection of the gray region. Unfortunately, the project budget is always limited, and so are the numbers of samples that may be collected.

While selecting the width of the gray region, the planning team evaluates the natural variability of the matrix to be sampled and its effects on the number of samples that may be collected. The lower the variability, the more accurately the sample mean concentration will represent the true mean concentration. Low variability data have higher confidence levels and smaller confidence intervals. (These concepts are defined in Appendix 1.) For matrices with high variability, the same may be achieved by collecting more samples, however this may not be feasible due to budget limitations. The planning team may base the gray region selection on the evaluation of risk associated with false acceptance decisions, rely on recommendations of regulatory guidance documents, or use professional judgment and experience with similar sites. The selection of the gray region for Case Study A is illustrated in the box on the next page.

2.1.6.5 Assign probability values to points above and below the action level

The planning team is now prepared to assign *decision error limits* to false acceptance and false rejection decision errors. *A decision error limit is the probability that an error may occur when making a decision based on sample data.* The probability curve tells us that the highest probability of error exists in the gray region; this error goes down as the mean concentrations move away from either side of the action level. The probability curve reflects our level of tolerance to uncertainty associated with a decision or, conversely, level of confidence with which a decision will be made.

The planning team will *select* and *state* the decision error limits based on such considerations as regulatory guidance; natural matrix variability; practical constraints (cost and schedule); physical constraints; potential impacts on human health and the

Step 6. Specify Limits on Decision Errors	
Case Study A	**Case Study B**
1. What is a range of possible values of the parameter of interest where the consequences of decision errors are relatively minor ('gray region')?	
Based on expected sample matrix variability, analytical method performance criteria, and budgetary considerations, the gray region is specified at 20 percent of the action level and ranges from 0.8 to 1.0 µg/kg.	Because of the non-probabilistic sampling approach, this step is not applicable. Several measures are proposed as a means to reduce the probability of discharging water with the VOC concentrations exceeding the discharge permit limitations, such as the following: • More frequent sampling • Specific QA/QC requirements to control the uncertainty associated with sampling error • More stringent method performance criteria to control analytical error

environment; or even political issues that may influence tolerance to decision error. Severe consequences will require more stringent decision error limits than moderate consequences. The selection of probabilities also depends on the expected distribution of the data (normal, lognormal, etc.); the type of the proposed sampling design; the number of samples to be collected; and the type of the statistical test that will be used for data evaluation.

Figure 2.3 shows the probability limits as *tolerable false acceptance and false rejection decision error rates*. In this figure, the *assigned* tolerable false acceptance decision error rate (β) increases from 0.05 to 0.10 as the true mean concentration increases and approaches the action level. This means that the planning team is willing to accept a 5–10 percent risk of making a false acceptance error. Inside the gray region (80–100 mg/kg), false acceptance error rates are not assigned and are considered tolerable, no matter how high they may be. On the other side of the action level, for the true mean concentrations greater than 120 mg/kg, the assigned tolerable false rejection decision error rate (α) is 0.01. This means that a risk of making an error in recognizing that the true mean concentration exceeds 100 mg/kg is only 1 percent. For the range of true mean concentrations from 100 to 120 mg/kg, the planning team assumes a higher tolerable false rejection decision error rate and consequently a higher level of risk (the probability of 0.1 or a 10 percent risk). The assigned probability values for the full range of concentrations in Figure 2.3 may be summarized as follows:

True mean concentration	*Tolerable probability of making an incorrect decision*
Below 70 mg/kg	$\beta=0.05$
70 to 80 mg/kg	$\beta=0.05–0.10$
80 to 100 mg/kg	Unassigned, gray region
100 to 120 mg/kg	$\alpha=0.10–0.01$
Greater than 120 mg/kg	$\alpha=0.01$

The planning team will use similar rationale for assigning tolerable decision error rates in an opposite scenario illustrated in Figure 2.4. In the planning phase, these probabilities will be factored into statistical tests for the calculation of the proper number of samples to be collected. In the assessment phase, they will be used in a statistical verification of the attainment of the action level.

The EPA recommends the most stringent decision error limits (0.01 for false acceptance and false rejection decision errors or $\alpha=\beta=0.01$) as a starting point in selecting the appropriate decision error limits (EPA, 2000a). The more practical limits are the less stringent ones: $\alpha=0.05$ for false rejection decision errors and $\beta=0.20$ for false acceptance decision errors (EPA, 1996b).

Step 6.　Specify Limits on Decision Errors	
Case Study A	**Case Study B**
1. What are tolerable decision error rates?	
The false acceptance and the false rejection decision error rates are assigned as follows: $\alpha = 0.10$; $\beta = 0.10$.	Because of the non-probabilistic sampling approach, this step is not applicable.

As we can see, Step 6 is the most complicated of all steps of the DQO process. The outputs of Step 6 are *the statement of baseline condition, the selection of the gray region, and the assigning of tolerable decision error rates.* Step 6 establishes the level of confidence with which decisions will be made and sets the stage for future statistical evaluation of the obtained data. Although Step 6 may not fully apply to some projects, it is important to understand the concepts it introduces as they are related to the concept of total error.

2.1.7　Step 7—Optimize the design for obtaining data
In Step 7 of the DQO process the project team will optimize the data collection design by performing the following actions:

1. Review the DQO outputs
2. Develop data collection design alternatives
3. Formulate mathematical expressions for each design
4. Select the number of samples that satisfies DQOs
5. Decide on the most resource-effective design or agreed alternative
6. Document the DQO outcome in the SAP

The planning team summarizes the output of the first six steps of the DQO process in a series of qualitative and quantitative statements that comprise the requirements for data collection and further adjusts them to meet the project objectives and budget. The team reviews the existing data and gathers the information that is useful for the new data collection design, such as the data on sample matrix interferences, or identifies the data gaps to be filled.

The team may revise the theoretical decision rule developed in Step 5. This rule was based on an assumption that the true mean contaminant of concern concentration is known and decisions will be made using unlimited data. In practice, the true mean concentration will be only approximated by sample mean concentration calculated from a limited number of samples. To be able to make decisions based on sample data, the team will create an *operational, statistically based decision rule* that may involve a statistical interval, such as confidence interval defined in Appendix 1. Using existing information or relying on prior experience with similar projects, the team will make statistical assumptions to derive the operational rule that is similar in intent to the theoretical rule. Using the outputs of Step 6 (possible range of the parameter of interest tolerable decision error rates), the planning team may apply statistical tests, such as ones recommended by the EPA (EPA, 1997a), to calculate the number of samples that must be collected to satisfy the DQOs. Typically, the planning team will invite a statistician who specializes in environmental chemical data evaluations to perform these calculations.

On occasion, however, even a non-statistician may be able to estimate a number of samples to be collected. Equation 11, Appendix 1, shows the calculation of the estimated number of samples that is used for hazardous waste characterization (EPA, 1996a). This equation assumes a normal distribution of individual concentrations and a limited number of samples (below 30). The calculation of the estimated number of samples is demonstrated in Example 2.3.

Example 2.3:   The calculation of the estimated number of samples

Excavated soil suspected of containing lead has been stockpiled. We may use this soil as backfill if the mean lead concentration is below the action level of 100 mg/kg. To decide if the soil is acceptable as backfill, we will sample the soil and analyze it for lead. The mean concentration of lead in soil will represent the statistical population parameter. The theoretical decision rule, the baseline and the alternative conditions, and the null and alternative hypotheses have been stated in Examples 2.1 and 2.2. The assigned probability limits are $\alpha = \beta = 0.05$. This means that the false acceptance error rate is 0.05. The probability of making a correct decision $(1-\beta)$ is 0.95 or the confidence level is 95 percent.

How many samples do we need to collect in order to establish that the mean concentration of lead is below the action level of 100 mg/kg with a confidence level of 95 percent?

Because historical data for the soil is not available, to estimate the number of samples for the stockpile characterization, we collect four preliminary samples and analyze them for lead. The concentrations are 120, 60, 200, and 180 mg/kg, with the average concentration of 140 mg/kg and the standard deviation of 63 mg/kg. We use Equation 11, Appendix 1, for the calculation of the estimated number of samples. Using a one-tailed confidence interval and a probability of 0.05, we determine the Student's 't' value of 2.353 for 3 degrees of freedom (the number of collected samples less one) from Table 1, Appendix 1.

The estimated number of soil samples is 14. For extra protection against poor preliminary estimate, we decide to collect 20 samples.

In Step 7, the planning team evaluates the data collection design alternatives and selects the most cost-effective approach that most closely meets the DQOs. Budget considerations always play a significant part during this step and often cause a change in the entire DQO process. To select the most cost-effective design, the team may have to relax the initially selected values for the gray region and for the probabilities of the tolerable decision error rates or choose to collect fewer samples, thereby decreasing the level of confidence.

Step 7. Optimize the Design for Obtaining Data	
Case Study A	**Case Study B**
1. What is the selected sampling design?	
The statistician reviews the outputs from the previous steps and determines that the DQOs will be best satisfied through the application of a statistical sampling design based on a systematic sampling from a hexagonal grid system. Six discrete samples will be collected from each grid unit followed by compositing into one sample for analysis.	Effluent will be sampled according to the frequency specified in the NPDES permit. In addition, influent will be sampled monthly to calculate the GAC vessel train loading. A more stringent and sensitive analytical method EPA Method 8260 will be used instead of EPA Method 624 required by the NPDES permit.
2. What were the key assumptions supporting the selected design?	
Compositing technique was selected because it is known to reduce variability in the data and the cost of analysis. Statistical testing of the data collected, which assumes an asymmetrical, non-normal distribution of the data from the entire area, is also proposed for the evaluation of the attainment of the action level.	The information on the GAC vessel train loading will be used for predicting a need for a vessel replacement. Preventive replacement of nearly spent GAC vessels will reduce a risk of the effluent exceeding the permit limitations.

The final outcome of the seventh step is *a data collection design that meets both the project DQOs and the project budget*. The team will document the outcome of the DQO process and the sampling design in a SAP or in a similar planning document. There are many different formats of documenting the DQO process outcome, from a simple tabular summary to an elaborate document complete with statistics and graphics. At a minimum, the sampling design should include the basic sampling and analysis requirements, such as the following:

- Action levels
- Sample matrices
- Number of samples
- Sample type (grab or composite)
- The locations of sampling points
- Sampling frequency
- Field quality control

- Sampling procedures
- Analytical methods
- Acceptance criteria for the PARCC parameters

2.1.8   DQO process in simple terms

Is there a need to conduct the DQO process for every environmental project? Not necessarily so. The DQO process can be lengthy and complicated. Because it is often poorly understood, it also can be somewhat intimidating. And yet, in essence, the DQO process addresses very basic issues that lay the foundation for the project's success. Systematic planning that forms the core of the DQO process is 'the scaffold upon which defensible site decisions are constructed' (Lesnik, 2001).

To conduct systematic planning for *any* environmental data collection project, we need to put Kipling's six honest serving-men to work, and make them find the answers to simple, yet fundamental questions:

- *What* is the project's purpose?
- *What* is the problem that requires data collection?
- *What* types of data are relevant for the project?
- *What* is the intended use of the data?
- *What* are the budget, schedule, and available resources?
- *What* decisions and actions will be based on the collected data?
- *What* are the consequences of a wrong decision?
- *What* are the action levels?
- *What* are the contaminants of concern and target analytes?
- *What* are the acceptance criteria for the PARCC parameters?
- *Who* are the decision-makers?
- *Who* will collect the data?
- *Why* do we need to collect the particular kind of data and not the other?
- *When* will we collect the data?
- *Where* will we collect the data?
- *How* will we collect the data?
- *How* will we determine whether we have collected a sufficient volume of data?
- *How* will we determine whether the collected data are valid?
- *How* will we determine whether the collected data are relevant?

Answering these questions and understanding the uncertainties associated with sampling and analysis is crucial for developing any data collection design. Several practical tips for conducting the DQO process are offered in the box on the next page.

2.1.9   Data quality indicators

The quality of analytical data is assessed in terms of *qualitative* and *quantitative data quality indicators*, which are *precision, accuracy, representativeness, comparability,* and *completeness* or the PARCC parameters. The PARCC parameters are the principal DQIs; the secondary DQIs are *sensitivity, recovery, memory effects, limit of quantitation, repeatability,* and *reproducibility* (EPA, 1998a).

Practical Tips: Data quality objectives

1. Understand and clearly state the purpose of data collection and the intended use of the data during project planning, whether it is conducted through the DQO process or through another systematic approach.

2. Identify target analytes and list them with their respective Chemical Abstract Service (CAS) Registry Number in the SAP to avoid confusion with different chemical names for the same compounds.

3. Select analytical methods appropriate for the intended use of the data by comparing the action levels to the method detection limits.

4. If the scale of decision-making (DQO Step 4) is not defined in the planning phase, some projects will be likely to exceed the budget due to unnecessary sampling and analysis.

5. Define in the planning documents the types of statistical tests that will be used for data evaluation.

6. Use one-tailed statistical test when comparing a data set to an action level to decide whether a mean concentration is above or below the action level.

7. Use two-tailed statistical tests when comparing two statistical parameters to each other.

8. Use Student's t-test for small, normally distributed populations (number of samples is less than 30).

9. Use standard normal distribution (z-statistic) for large populations (number of samples is greater than 30).

10. A 90 percent confidence level and a 10 percent error rate will be appropriate for most projects where mean concentrations are compared to action levels.

11. The EPA recommends an 80 percent confidence level and a 20 percent error rate when comparing the mean concentrations to regulatory levels for decisions related to determining whether a waste is hazardous or not.

12. Involve a statistician into the planning process for sites with highly toxic chemicals or with stratified contaminant distribution for the determination of background concentrations or for 'hot spot' detection projects.

13. Do not overlook the DQOs for project-generated waste characterization and disposal, as they are different from these for the main set of data.

Qualitative and quantitative acceptance criteria for the PARCC parameters are derived in the planning phase. Whether they are specific statistical values or represent accepted standards and practices, they must be always selected based on the project objectives and be appropriate for the intended use of the data. The DQI acceptance criteria are documented in the SAP and serve as standards for evaluating data quality and quantity in the assessment phase of data collection process. The primary DQIs are established through the analysis of field and laboratory QC samples and by adhering to accepted standards for sampling and analysis.

DQIs are usually thought of as attributes of a laboratory measurement system. However, a broader definition of primary DQIs will enable us to assess the entire measurement system that includes not only the laboratory measurements but also the sampling design and field procedures. Such broad interpretation of the primary DQIs will allow us to evaluate all components of total error and with it the overall, not just the analytical, data quality. The DQI definitions (EPA, 1999a) presented in this chapter are interpreted in a manner that encompasses all qualitative and quantitative components of total error.

2.1.9.1 Precision
Precision is the agreement between the measurements of the same property under a given set of conditions. Precision or random error is a quantitative parameter that can be calculated in several different ways: as the standard deviation; relative standard deviation (RSD); or as relative percent difference (RPD). The first two are common statistical parameters that are used for the evaluation of multiple replicate measurements, whereas RPD is used for measuring precision between two duplicate measurements. Equation 1 in Table 2.2 illustrates the method for calculating RPD as a measure of precision.

Intralaboratory analytical precision is the agreement between repeated measurements of random aliquots of the same sample made at a single laboratory using the same method. Interlaboratory precision is the agreement between the measurements of the same sample conducted by two or more laboratories. Intralaboratory and interlaboratory precision may be calculated as the RPD for two measurements, as variance or as the standard deviation for more than two measurements.

Laboratories establish analytical precision for each method using a *laboratory control sample* (LCS) and *laboratory control sample duplicate* (LCSD). These samples are made at the laboratory with interference-free matrices fortified (spiked) with known amounts of target analytes. Interference-free matrices are analyte-free reagent water or laboratory-grade (Ottawa) sand. Precision is then calculated as the RPD between the results of the LCS and LCSD. Analytical precision depends on analytical method and procedure; the nature of the analyte and its concentration in the LCS and LCSD; and the skill of the chemist performing analysis. The RPD for interference-free laboratory QC samples is typically below 20 percent for soil and water matrices.

The RPD calculated for pairs of identical environmental samples (field duplicates) is the measure of total sampling and analysis precision, which combines the precision of sampling, sample handling, and the precision of sample preparation and analysis. Precision of field duplicates may be significantly affected by matrix interferences and by inherent sample variability. That is why the SAP should make a distinction between analytical precision determined from LCS/LCSD pairs and total sampling and analysis precision determined from field duplicate pairs and adopt separate acceptance criteria for each.

2.1.9.2 Accuracy
Accuracy measures the closeness of a measurement to the true parameter value. Accuracy is a combination of random error (precision) and systematic error (bias). A data set that has a high precision and a high systematical error (high bias) does not

A practical way to calculate completeness is on an individual method basis. *Individual method completeness is the number of valid measurements as a percentage of the total number of measurements in one analytical method.* A single parameter analysis may be invalid for all samples, but it will not have much effect on the completeness calculation if merged with a large total analyte number. This one method may, however, be of critical importance for the project decisions. Calculating method analytical completeness enables us to determine whether any of the performed analytical methods fail to provide a sufficient quantity of valid data and, consequently, whether resampling and reanalysis may be needed.

Sampling completeness

Completeness is a measure of whether all of the planned samples have been collected. This is sampling completeness, which is calculated according to Equation 3c, Table 2.2. Sampling completeness is important for projects, which are planned on accurate predictions of the number of samples required to fulfill the DQOs or which involve a statistical test as the final data use. For such projects, the lack of sampling completeness or analytical completeness alike will decrease the power of the statistical test and affect the error rates. However, no matter how good our planning may be, unforeseen factors, such as refusal during drilling, dry ground-water wells, inaccessible sampling points, may preclude us from collecting samples as planned. Another source of poor sampling completeness are the collected samples, which due to errors in sampling or sample handling became not viable for analysis.

For many projects, sampling completeness may not be important and cannot be calculated. In fact, some projects (for example, removal actions with unknown spatial extent of contamination, unknown waste profiling, process monitoring, characterization of contaminated construction debris during building demolition) are not conducive for predicting the exact number of samples that will be collected. That is why during the planning phase of any project we should evaluate the importance of sampling completeness and decide if its calculation will enhance the project data quality.

Holding time completeness

The holding time completeness is the number of analyses conducted within the prescribed holding time as a percentage of the total number of analyses. It is calculated using Equation 3d, Table 2.2. *Holding time is the period of time a sample may be stored to its required analysis* (EPA, 1998a). The importance of the holding time of analysis is discussed in Chapter 3.

The holding time observance is the first of the criteria, which we use to judge whether the data are valid or invalid. Expired samples (samples analyzed past the holding time) may produce invalid results and may require resampling and reanalysis. The acceptance criteria goal for the holding time completeness should be always 100 percent, for holding time is a primary indicator of our ability to manage projects in the field and at the laboratory.

Secondary data quality indicators

The secondary DQIs are not immediately obvious to the data user; not all of them are applied for data quality evaluation. Nevertheless, they are among the fundamental concepts of analytical chemistry, have a great effect on results of qualitative and quantitative analysis, and affect the outcome of the primary DQIs. The meaning and importance of the secondary DQIs are discussed in Chapter 4, which details laboratory analysis and quality control. The secondary DQIs, which include sensitivity, recovery, memory effects, method detection limit, limit of quantitation, repeatability, and reproducibility, are defined as follows:

- *Sensitivity* is the capability of a method or instrument to discriminate between measurement responses for different concentration levels (EPA, 1998a). It is determined as a standard deviation at the concentration level of interest and, therefore, it is concentration-dependent. It represents the minimum difference between two sample concentrations with a high degree of confidence. Laboratories establish a method's sensitivity by determining the *method detection limit* (MDL). The MDL is the minimum concentration of a substance that can be measured and reported with 99 percent confidence that the analyte concentration is greater than zero (EPA, 1984a).

- *Recovery* is an indicator of bias in a measurement (EPA, 1998a). It is evaluated by measuring the reference materials or samples of known composition. Environmental analytical methods establish recovery based on additions of surrogate standards or target analytes to environmental and laboratory QC check samples. The recovery is calculated as a percentage of a measured concentration relative to the added concentration (Equation 2, Table 2.2). Unbiased measurements have a 100 percent recovery. As a means to establish a range of acceptable recoveries, the laboratories maintain control charts (plotting the recoveries and calculating control limits as a function of the standard deviation). Exceedingly low or high recoveries are indicators of analytical problems.

- *Memory effects* or carryover occur during analysis when a low concentration sample is influenced by a preceding high-concentration sample (EPA, 1998a). Memory effects take place because of contamination of the analytical system originating from high contaminant concentrations in a sample. As a result, the accuracy of the measurements of subsequent samples may have a high bias. Memory effects are a common cause for false positive results in laboratory analysis.

- *The limit of quantitation* is the minimum concentration of an analyte in a specific matrix that can be determined above the method detection limit and within specified bias and precision limits under routine operating conditions (EPA, 1998a). The limit of quantitation is often referred to as the *practical quantitation limit* (PQL). The concept of PQL is discussed in Chapter 4.5.1.

- *Repeatability* (intralaboratory precision) is the degree of agreement between independent test results produced by the same analyst using the same test method and equipment on random aliquots of the same sample within a short time period (EPA, 1998a). Repeatability is calculated as the RPD for two measurements, as the variance or the standard deviation for more than two measurements.

- *Reproducibility* (interlaboratory precision) is the precision that measures variability among the results obtained for the same sample at different laboratories (EPA, 1998a). It is usually statistically expressed as variance or the standard deviation. The low variance or the standard deviation value indicates a high degree of reproducibility.

2.2 DEFINITIVE, SCREENING, AND EFFECTIVE DATA

Environmental data may be classified according to the level of associated total error as *screening* data and *definitive* data. These two data types are also different with respect to their intended use and QC elements. The selection of the type of data to collect depends on the project DQOs and is made in the planning phase. The realm of possible data uses and methods of data collection are the defining factors in the selection.

Definitive data are obtained with rigorous analytical methods, such as EPA-approved methods or other standard analytical methods. For the data to be definitive, either analytical or total measurement error must be determined. Definitive data, which are analyte-specific and have a high degree of confidence in analyte identity and concentration, are used for decisions that have consequences for human health and the environment, such as site closure, risk assessment, and compliance monitoring of water effluents and air emissions. Definitive data may be generated at a field (mobile) laboratory or at an off-site (fixed-base) laboratory.

Screening data have a different purpose and application. Screening data enable us to obtain as much information as possible within the shortest period of time and in the most inexpensive manner. Compared to definitive data, screening data are generated by rapid, less precise methods of analysis with less rigorous sample preparation. For some screening methods, total error or even measurement error cannot be evaluated due to the nature of the employed detection technique. Screening data are not necessarily analyte-specific and may represent groups of compounds of a similar chemical or physical nature. Although screening data quantitation may be relatively imprecise, the data should nevertheless be of known quality. It means that screening methods of analysis should have a defined set of QC requirements and QA procedures. To further establish the level of screening data reliability, they are confirmed by definitive laboratory analysis, typically at a 10 percent rate of all collected samples.

The use of screening data enables project teams to estimate an environmental condition in a rapid manner and to facilitate real-time decision-making in the field. Screening results usually determine future action at the project site, and that is why screening data, imprecise as they may be, must reliably reflect the true site conditions. Screening data may help formulate important decisions for many types of environmental projects, such as the following:

- Placement of groundwater monitoring wells
- Contaminant plume delineation
- Soil characterization during site assessment
- Excavation guidance
- Treatment system startup and monitoring

• Field measurements of groundwater parameters

A great variety of screening methods and tools exist today; new techniques based on the latest scientific advances are being rapidly developed. Among established screening methods are immunoassay and colorimetry tests for a vast range of contaminants, and X-ray fluorescence (XRF) screening for metals. The use of field mobile analytical laboratories is an excellent way to obtain screening data, since laboratory methods for definitive data analysis can be easily adapted for fast screening data generation at a project site.

When planning screening data collection, we should always perform a cost-benefit analysis, i.e. compare the cost of screening to the volume and the value of the obtained information. The cost of screening, including field kits, instruments, operator's labor or mobile laboratory rates should be compared to the availability and cost of laboratory analysis. The reliability of a screening technique should be verified first to establish whether it would bring the desired results, since matrix interferences or the degradation of contaminants of concern often renders a screening method unusable for a specific application. Only after having evaluated all of these issues, we should be able to make a decision on the use of field screening methods.

There is no distinct difference between certain types of screening data and definitive data, and definitive data are always subject to risk of becoming screening data. For example, definitive data obtained with standard analytical protocols at a fixed-based laboratory may be downgraded to screening data due to deficiencies in data quality or the overpowering effects of matrix interferences. As a consequence, the use of this data set may be different from that originally intended. On the other hand, screening data due to their inherent uncertainty cannot become definitive under any circumstances.

Screening data are often compared to the action levels for the purpose of making preliminary decisions during project implementation. For final decision-making, only definitive data of acceptable quality should be used for comparisons to the action levels.

Understanding the nature of the environmental sample and the limitations it places on accuracy and precision of analysis is a critical factor in selecting what type of data to collect. Some matrices are so inherently indefinite that the analytical data produced for them even with the most accurate and precise methods cannot be called definitive, as illustrated in Example 2.4.

Example 2.4:   Matrix limitations

Soil gas surveys are widely used for delineating sites with VOC contamination. Even though soil gas data may be generated with the most definitive and accurate analytical method available, such as gas chromatography/mass spectrometry (GC/MS), they will always remain screening data due to uncertainty associated with the sample matrix.

Oily matrices (sludge, fuel-contaminated soil) may undergo multiple dilutions during analysis. The interferences from the oil constituents and the analytical uncertainty induced by dilutions may reduce the data obtained with definitive analytical methods to screening data.

- *Listed* wastes are hazardous waste streams from specific and non-specific sources, listed under codes F, K, P, and U in 40 Code of Federal Regulations (CFR) Part 261.4
- Hazardous debris

The most commonly used action levels for determining whether a waste stream exhibits toxicity characteristic and therefore is hazardous under RCRA are the Toxicity Characteristics Leaching Procedure (TCLP) concentrations for 24 organic compounds, 6 pesticides, 2 herbicides, and 8 metals as described in 40 CFR Part 302. Appendix 3 presents a list of these compounds and their maximum TCLP concentrations, which are frequently referred to in project work.

The RCRA treatment standards are available from 40 CFR Parts 261, 264, 265, 268, 271, and 302. Numerous reference guides and manuals on handling of RCRA wastes are available to environmental professionals for educated decision-making. However, one should keep in mind that hazardous waste disposal is an extremely complex process and that it is best served by hazardous waste and regulatory compliance experts who have a detailed knowledge of the applicable laws.

In California Code of Regulations (CCR) Title 22, the State of California imposes additional requirements for the disposal of waste containing 20 inorganic and 18 organic persistent and bioaccumulative toxic substances (CCR, 1991). Hazardous characteristics of waste streams contaminated with these substances are determined as Total Threshold Limit Concentrations (TTLC) and Soluble Threshold Limit Concentrations (STLC), shown in Appendix 4.

Site investigation and remediation projects usually include the disposal of investigation-derived waste (IDW), for example, soil from cuttings or excavations, equipment decontamination water, purged groundwater from well installation, etc. To determine whether these waste streams may be hazardous, we should consider their source, and, if necessary, chemically characterize the streams for the determination of disposal options. The disposal facility acceptance criteria would be the action levels in this case.

2.3.6 Polychlorinated biphenyl cleanup levels

Media contaminated with polychlorinated biphenyls (PCBs) are regulated by the Toxic Substances Control Act (TSCA) signed into law in 1976. The TSCA also provides standards for asbestos, chlorofluorocarbons (CFCs), and lead, and enables the EPA to track a total of 75,000 industrial chemicals currently produced or imported into the USA.

The TSCA requirements for the disposal of PCB-contaminated waste and PCB-containing material can be found in 40 CFR Parts 750 and 761 (EPA, 1998c). The disposal of waste streams containing PCBs at concentrations exceeding 50 parts per million (ppm) are regulated under the TSCA. For the purposes of cleaning, decontaminating, or removing PCB waste, the TSCA defines four general waste categories: bulk PCB remediation waste, non-porous surfaces, porous surfaces, and liquids. Each type of waste has a regulatory cleanup level shown in Appendix 5.

In 40 CRF Section 761 Subpart O, the TSCA specifically addresses the requirements for verification of self-implementing cleanup by defining the numbers and locations of samples to be collected from bulk PCB remediation waste and porous surfaces. Section 761 Subpart P specifies the requirements for non-porous surfaces cleanup verification. These regulation-prescribed sampling designs are discussed in detail in Chapter 3.5.2.

2.4 HOW TO NAVIGATE THE ANALYTICAL METHOD MAZE

An analytical method is a procedure used for determining the identity and the concentration of a target analyte in a sample. All analytical methods used in environmental chemical pollutant data collection may be categorized as *regulatory* or *consensus*. The former are approved by the EPA for mandatory use under a program or a law; the latter are developed and published by professional organizations on a consensus basis and do not need the EPA approval. Published analytical methods, regulatory or consensus alike, typically include the information on sample collection, preservation, storage, and holding time; sample preparation and analysis; compound identification and quantitation; and analytical quality control.

The number of existing methods for the analysis of environmental pollutants is staggering. Considering how many of them are redundant or obsolete and how many of them are designed to analyze for the same chemicals, one cannot help but question their necessity. The method maze owes its existence to the progress of environmental protection in the USA, and the emergence of different analytical methods is directly related to the introduction of various environmental laws. The evolution of methods themselves is linked to technological progress, and every new analytical method reflects the state-of-the-art that existed at the time when the method was first introduced. The following brief chronological summary of existing test methods sheds light on their history and applicability today.

2.4.1 Consensus methods

Standard Methods for the Examination of Water and Wastewater, first published in 1905, is a manual that can be considered the mother of all methods. It is currently prepared and published jointly by the American Public Works Association (APHA), American Water Works Association, and Water Environment Federation, and had undergone its 20th edition in 1998. This extensive manual contains hundreds of test methods for the analysis of domestic and industrial water supplies, groundwater, surface waters, process water, and treated and untreated municipal and industrial discharges. Some of the methods from this manual were adopted into the EPA method framework, for example, the methods for trace element and individual and aggregate (non-individual, combined) organic compound analyses, while others, such as microbiological testing, do not have any EPA counterparts. This manual is not directly linked to any existing laws. Its great value is in the solid technical foundation and QC protocols. In environmental project work, we commonly use the manual's methods for microbiology and the determinations of selected physical and chemical properties.

American Society for Testing and Materials (ASTM) *Annual Book of Standards* is another important source of consensus methods used in environmental project work.

The EPA approved many of the methods published in the *Annual Book of ASTM Standards* for compliance monitoring under the CWA. ASTM methods used in environmental remediation projects determine soil properties, such as organic content, porosity, permeability, soil grain size, or the properties of free petroleum product contained in the subsurface, such as viscosity, density, and specific gravity.

The use of American Petroleum Institute (API) Recommended Practices is not uncommon in environmental project work, particularly for the characterization of recovered petroleum products that have accumulated in the subsurface from leaking USTs and pipelines.

2.4.2   Methods for water compliance monitoring

Methods for Chemical Analysis of Water and Wastes, EPA-600/4–79–020, first published in 1979, is a manual of test methods approved for the monitoring of water supplies, waste discharges, and ambient waters under the SDWA and the NPDES. These methods, numbered as the *100, 200, 300*, and *400 Series*, include the testing of physical parameters, trace elements, inorganic non-metallic parameters, and aggregate organic parameter (EPA, 1983). Today they are *mandatory* for the analysis of groundwater, surface waters, domestic and industrial sludge effluents, and treatment process water; and are widely conducted as required by the NPDES permits.

Guidelines for Establishing Test Procedures for the Analysis of Pollutants under the Clean Water Act, 40 CFR Part 136, were first promulgated in 1973 and amended in 1984 to include additional methods. These test procedures are *mandatory* for CWA compliance monitoring and for wastewater pollutants in municipal and industrial wastewater discharges under the NPDES permits (EPA, 1984b). The list of pollutants includes biological, inorganic, non-pesticide organic, pesticide, and radiochemistry parameters. The organic analysis series for VOCs, semivolatile organic compounds (SVOCs), and pesticides include the so-called *600* and *1600 Series* methods. Biological and inorganic tests for CWA monitoring default to the APHA *Standard Methods* and *Methods for Chemical Analysis for Water and Wastes* (EPA, 1983); radiochemistry methods are from *Prescribed Procedures for Measurement of Radioactivity in Drinking Water* (EPA, 1980). The new 1600 Series methods added to 40 CFR Part 136 can be found on the EPA's Office of Water Web page at http://www.epa.gov/ost/methods.

In recent years, many of the technologically outdated methods of 40 CFR Part 136 have been upgraded to incorporate the latest advances in instrumental analysis. For example, capillary chromatographic columns with superior compound resolution replaced obsolete packed columns in gas chromatography (GC) and GC/MS analytical methods; Freon 113, a chlorofluorocarbon harmful to the environment, was phased out as the extraction solvent in oil and grease analysis and replaced with hexane in Method 1664 (EPA, 1999b).

Another example of the emergence of a new, technology-driven method is the analysis of mercury by oxidation, purge and trap, and cold vapor atomic fluorescence spectrometry by EPA Method 1631 (EPA, 1998d). The advantage of this method is in its low detection limit of 0.5 nanograms per liter (ng/l) or 0.5 parts per trillion. The detection of mercury at such a low concentration is susceptible to false positive results due to contamination originating from sampling, sample handling, and analysis itself.

To prevent trace level contamination, the EPA established special sampling techniques for trace elements in ambient water in Method 1669 (EPA, 1996c). The EPA also provided the guidance for clean room laboratory analysis and for documentation and evaluation of trace element data (EPA, 1996d; EPA, 1996e).

2.4.3   Methods for drinking water analysis

After SDWA had established the MCLs as the standards for drinking water and groundwater quality, the EPA developed a series of analytical methods, known as the *500 Series*, specifically for the determination of trace level organic compounds in interference-free aqueous matrix, i.e. drinking water. These *mandatory* methods were introduced in the *Methods for the Determination of Organic Compounds in Drinking Water*, first published in 1988, revised in 1991, and further supplemented in 1992 (EPA, 1992). In addition to organic pollutants, the SDWA regulates inorganic chemicals and secondary water quality parameters, radionuclides, disinfectant residuals, and microbiological contaminants. For the analyses of non-organic parameters, the SWDA defaults to the ASTM's *Annual Book of Standards, APHA Standard Methods*, the EPA 100 to 400 Series, and other methods developed by private industry or consensus organizations.

A list of analytical methods approved for the SDWA compliance monitoring can be found on the EPA's Office of Groundwater and Drinking Water Web page at http://www.epa.gov/ogwdw/methods.

The EPA or the state certifies analytical laboratories, which perform drinking water analysis according to the requirements of these methods. The laboratories must participate in the EPA PE sample program and undergo periodic state audits. In some states, all groundwater is analyzed using drinking water methods (for example, Arizona, Utah), whereas in others, depending on whether or not the groundwater may be used as a source of drinking water, non-drinking water methods may be used.

2.4.4   Methods for hazardous waste analysis

Test Methods for Evaluating Solid Waste, SW-846, Physical/Chemical Methods, first published in 1986, is a living document that underwent its fourth update in 2001. This manual and its methods are summarily referred to as the *SW-846* or the *RCRA methods*. This two-volume, 13-chapter manual provides the *guidance* (not mandatory) test procedures approved by the EPA's Office of Solid Waste for obtaining data that satisfy the RCRA requirements of 40 CFR Parts 122 through 270.

The manual presents the state-of-the-art analytical testing of trace element and organic and inorganic constituents in various matrices that may be sampled and analyzed to determine compliance with the RCRA regulations. The samples physical states include aqueous and oily sludge, TCLP extract, soil, groundwater, stack condensate, and some others. SW-846 also describes procedures for field and laboratory quality control and field sampling. It contains methods for determining the physical properties of wastes and their hazardous characteristics (toxicity, ignitability, reactivity, and corrosivity). SW-846 Chapter Two provides guidance on the selection of appropriate methods depending on the matrix and the target analytes. The SW-846 methods can be found on the Office of Solid Waste and Emergency Response Web page at http://www.epa.gov/epaoswer/hazwaste/test/sw846.htm.

SW-846 Volume I contains the preparation and analysis methods divided into series according the nature of the tested parameters:

- 1000 Series—Hazardous characteristics of wastes
- 3000 Series—Sample preparation and cleanup methods for metal and organic compound analysis
- 4000 Series—Immunoassay screening methods
- 5000 Series—Sample preparation for VOC analysis
- 6000 and 7000 Series—Trace element analysis
- 8000 Series—Individual organic compound analysis
- 9000 Series—Miscellaneous tests (aggregate organic compound analysis, anions, coliform, radiochemistry)

SW-846 Volume II, *Field Manual*, details statistical and non-statistical aspects of sampling and provides information on the regulatory issues for several monitoring categories, such as groundwater monitoring (Chapter 11), land treatment (Chapter 12), and incineration (Chapter 13). The purpose of this guidance is to assist the user in the development of data quality objectives, sampling and analysis plans, and standard operating procedures (SOPs).

When published as a new method of SW-846, a method's number, for example, 8270 or 8015, does not include a letter suffix. However, each time the method is revised and promulgated as part of an SW-846 update, it receives a letter suffix. For example, a suffix 'A' indicates the first revision of that method; a suffix 'B' indicates the second revision, etc. Under Update III, Method 8270 became 8270C, reflecting a third revision, and Method 8015 became 8015B, indicating a second revision. A method's reference found within the RCRA regulations and in the SW-846 text always refers to the latest promulgated revision of the method, even if the method number does not include the appropriate suffix. *To properly reference the SW-846 methods in the planning documents and during analysis, the entire method number including the suffix letter designation must be identified.*

The SW-846 methods are used in the RCRA compliance and monitoring programs for hazardous waste and stack gas characterization, groundwater monitoring, soil, surface water, and groundwater analysis during site investigation and remediation. Many laboratories adopted the SW-846 technical provisions as standards for routine operations. The fact that many of the SW-846 methods do not specify the acceptance criteria for analytical accuracy and precision and allow laboratories to use their own makes these methods even more attractive to the laboratories.

2.4.5 Contract Laboratory Program
Abandoned or inactive sites in the USA are regulated through the Comprehensive Environmental Response, Compensation, and Liability Act (CERCLA) or the Superfund, first signed in 1980 and reauthorized in 1986 in the Superfund Amendment and Reauthorization Act (SARA). *Superfund Contract Laboratory Program* (CLP) is a national network of EPA representatives, EPA contractors, and commercial environmental laboratories that support EPA's Superfund projects under

the CERCLA. Under the CLP, Analytical Operations Center (AOC) of the EPA's Office of Emergency and Remedial Response offers analytical services to Superfund project decision-makers. Samples of soil, sediment, and groundwater from Superfund hazardous waste sites, collected by the EPA's subcontractors for the Superfund site investigation, remediation, monitoring, or enforcement, are analyzed at the CLP laboratories. The EPA pays for sampling, analysis, and data evaluation using the money from the Superfund accounts. The AOC provides administrative, financial, and technical management of the CLP including on-site laboratory evaluations; the preparation and evaluation of PE samples; QA audits of the CLP data; and the development of new analytical methods. The EPA manages the sampling and analysis activities in a centralized manner through Regional Sample Control Coordinators at each of its ten Regions. The Regional Coordinators schedule all analytical work and track samples by the number, matrix, and analysis. The Regional Technical Project Officers monitor the CLP laboratories in their Regions, identify and resolve problems in laboratory operations; and participate in laboratory audits.

The CLP Statements of Work (SOWs) for organic and inorganic analyses provide detailed instructions for sample management, analysis, laboratory QA/QC procedures, and data reporting (EPA, 1999d; EPA, 2001). CLP data are evaluated (validated) using the acceptance criteria of the *USEPA National Functional Guidelines for Inorganic Data Review* (EPA, 1994) and *USEPA National Functional Guidelines for Organic Data Review* (EPA, 1999c). The use of standardized procedures in sample handling, analysis, and validation enables the CLP laboratories to produce data of known and documented quality in a consistent manner.

The CLP Target Compound List (TCL) includes 33 VOCs; 64 SVOCs; 28 organochlorine pesticides and PCBs. The Target Analyte List (TAL) consists of 23 trace elements and cyanide. A separate SOW exists for dioxin/furans. The high and low detection limits for soil and water matrices specified in the methods are a contract requirement and they give rise to the term of the Contract Required Quantitation Limits (CRQLs). Over the years, the CLP SOWs for Organic and Inorganic Analytical Services have undergone several revisions, which can be found on the EPA Superfund Web page at http://www.epa.gov/oerrpage/superfund/programs/clp/analytic.htm.

The benefits of the CLP are numerous; the fact that it is a turnkey program with an independent QA oversight provided by the EPA being the most important one. The observance of the CLP analytical protocols as a contract requirement ensures that the data obtained at different CLP laboratories are comparable. Extensive documentation makes data tracking and reconstruction possible; negotiated prices for laboratory contracts and the economy of scale due to large numbers of samples make the CLP cost-effective and efficient.

Over the years the CLP requirements greatly influenced the industry standard of sample handling, analysis, reporting, and data evaluation. Some of the CLP sample management procedures have been adopted as standard practices, for example, sample tracking procedures within the laboratory or grouping samples into sets known as Sample Delivery Groups (SDGs). In the absence of other guidance, the CLP National Functional Guidelines became the widely-used standards for data evaluation, despite the fact that their acceptance criteria do not apply to any sampling strategy must provide the data that accurately represent the site conditions, are appropriate for the intended use, and meet the DQOs.

2.5.1   Grab and composite samples

There are two types of samples that can be collected in a sampling event: ***grab*** samples and ***composite*** samples.

Grab samples are discrete aliquots representing a specific location at a specific time (DOE, 1996). The terms *grab* and *discrete* are used interchangeably for describing this type of samples (USACE, 1994). Chemical data obtained from grab samples represent an environmental condition at a specific location at the time of sampling. Grab samples are not combined (composited) with other samples. Typical grab samples are groundwater samples and soil samples collected for VOC analysis; samples collected from different depths in a soil boring; groundwater samples collected from different depths of a screened interval within a well.

Composite samples are samples composited of two or more specific aliquots (discrete samples) collected at various sampling locations and/or at different times (USACE, 1994). Composite samples are prepared by rigorous mixing of equal aliquots of grab samples. Analytical results from a composite sample may represent an average concentration for the points where grab samples were collected or an average concentration for the time period during which the grab samples were collected. Sample compositing does not adversely affect the method detection limits, which depend only on the weight of soil or the volume of water used for analysis.

Composite samples are typically collected for soil stockpile or surface area characterization. Twenty-four-hour composite samples of water may be collected with automated composite samplers from streams or process piping; composite samples may be made of several grab samples collected from different depths in a soil boring.

A properly prepared composite sample is also a ***homogenized*** one. *Samples collected for volatile contaminant analysis are never composited or homogenized*. This includes halogenated and aromatic solvents, gasoline, and other petroleum mixtures that contain volatile hydrocarbons with the carbon number below C_{15}. Because samples collected for VOC analysis cannot be homogenized, we often observe poor precision for volatile contaminant analyses. The effects of sample variability are often so substantial that a soil sample in a 6-inch core barrel liner, when subsampled for analysis from two different ends, may exhibit the VOC concentrations that are orders of magnitude different from each other. The same is often true for SVOCs and inorganic contaminants. Although homogenization is not possible for samples collected for VOC analysis, it is very important in SVOC and inorganic pollutant analysis, as it reduces sample variability and improves precision.

2.5.2   Probabilistic sampling

Probabilistic sampling, which lies in the core of the DQO process, is based on a random sample location selection strategy. It produces data that can be statistically evaluated in terms of the population mean, variance, standard deviation, standard error, confidence interval, and other statistical parameters. The EPA provides detailed guidance on the DQO process application for the development of probabilistic sampling designs under the CERCLA and RCRA Corrective Action Program (EPA, 2000b) and for sampling of hazardous waste (EPA, 1996a). Commonly used statistical parameters and calculations, which are part of the SW-846 Volume II, Field Manual (EPA, 1996a), are reproduced in Appendix 1.

Depending on the variability of the sampled medium and the purpose of sampling, several types of probabilistic sampling designs may be used.

Simple random sampling is used when the variability of the sampled medium is insignificant, and the contaminant may be assumed to be evenly distributed. Every possible sampling point has an equal probability of being selected. Each sampling point is selected independently from all other points, most often with the help of a random number generator program. Generally, simple random sampling is used when little information is available for a site or a waste stream. Simple random sampling is most appropriate for the characterization of relatively homogenous waste in holding vessels and lagoons or for batches of process waste (ash, sludge). For such media, one grab sample will represent the entire waste stream.

Stratified random sampling, which is a variation of simple random sampling, is used for media that are stratified with respect to their chemical and physical properties. Each stratum is identified and randomly sampled. The number of grab samples and the sampling point selection depend on the nature of contaminant distribution within each stratum. Stratified random sampling is used for the characterization of multiphase liquid wastes or process waste batches that undergo stratification over time and/or space.

Composite simple random sampling is appropriate for a medium with high variability. Grab samples are collected at random sampling points and mixed together into a composite sample prior to analysis. The advantage of this type of sampling is that a greater number of samples may be collected to better represent the mean contaminant concentration of the sampled medium without an increase in the analytical cost. That is why composite sampling is usually used as a cost-saving measure. To be able to obtain representative results from composite sample analyses, compositing must be conducted in a manner that ensures equal representation of every grab sample. Equal weights of grab samples must be thoroughly mixed (homogenized) to make a representative composite sample. Homogenization is particularly important for metal analysis of soil samples, because of a small analytical aliquot size (1 to 2 grams).

Systematic sampling consists of overlaying a two-dimensional or a three-dimensional grid over the area to be sampled followed by sampling from each grid unit. The starting point of the grid is randomly selected, and that is why this sampling strategy is often referred to as **systematic random sampling**. The orientation, the overall size of the grid, and the grid unit length (the distance between the sampling points) are based on the objectives of the data collection and are often designed by a statistician. Grab samples, which are collected from the nodules of a grid system, form a symmetrical square, rectangular, or hexagonal (triangular) pattern.

Systematic sampling on a two-dimensional grid is used for characterizing large areas of surface soil; floors and walls of excavation pits and trenches; and for the verification of decontaminated surfaces cleanup. An example of a systematic sampling on a three-dimensional grid is the sampling of a conical or elongated soil stockpile, divided into sections of equal volume. Each section is then sampled by collecting a certain number of random grab samples, which can be analyzed individually or composited prior to analysis.

2.5.3 Judgmental sampling

Contrary to probabilistic sampling, judgmental sampling is intentionally non-random and even biased. Data obtained through judgmental sampling cannot be statistically evaluated.

Judgmental sampling is used to obtain information about specific areas suspected of contamination. The sampling point selection is based on the existing information on the contamination source and history. Typically, an individual who has the detailed knowledge of the area to be sampled makes decisions on the sampling point placement.

This type of sampling may also rely on immediate observations. For example, judgmental sampling is appropriate for investigating hot spots, such as stains, spills, or discoloration, which may indicate contamination. The knowledge of the operational modes or past practices at the project site may also serve as a basis for judgmental sampling.

A variation of judgmental sampling is regulation-prescribed sampling during UST removal. Sampling points are predetermined based on predictable leakage or spilling points, such as the fill or the pump end of the UST or the joints in a pipeline. The number of samples typically depends on the size of the UST and the presence of groundwater in the tank pit. Prescriptive sampling for UST removal is implemented in several states in the USA.

2.6 FIELD QUALITY CONTROL AND QUALITY ASSURANCE SAMPLES

The collection of field QA/QC samples prescribed by the EPA and DOD guidance documents is part of every sampling event. Typically, QA/QC samples are collected to satisfy the protocol requirements and the obtained data are rarely used for project decisions. And yet, field QA/QC samples, like all other samples collected for the project, must have a well-defined need, use, and purpose and be relevant to the project objectives.

The following *field QC samples* may be collected during a sampling event and analyzed together with the field samples:

- Trip (travel) blanks
- Field duplicates
- Equipment (rinsate) blanks
- Temperature blanks
- Ambient (field) blanks

Field QA samples are replicates or splits of field samples that are analyzed at a different laboratory to establish data comparability. QA field samples may have their own set of associated field QC samples.

Table 2.4 summarizes existing practices for the collection and analysis of the field QC and QA samples and their practical value.

The current practice of QA/QC sample collection has a tendency for going to extremes without adding value to the project data. For example, the EPA recommends the following rates for their collection and analysis (EPA, 1996a):

- Field duplicate (one per day per matrix type)
- Equipment rinsate (one per day per matrix type)
- Trip blank (one per day, VOC analysis only)
- Matrix spike (one per 20 samples of each matrix type)
- Matrix spike duplicate (one per 20 samples of each matrix type)

The indiscriminate collection of QA/QC samples according to these guidelines may be too excessive for some projects and insufficient for the others. *The only appropriate vehicle for arriving at the proper type and quantity of QA/QC samples is the DQO process.* If the selection of the QA/QC sample type is not based on the project DQOs and their data are not meaningfully interpreted, their only contribution to the project is the increase in the cost of sampling and analysis.

2.6.1 Trip blanks

Trip blanks are QC samples associated with the field samples collected for VOC analysis, such as chlorinated and aromatic hydrocarbons and volatile petroleum products. Trip blanks are sometimes called *travel blanks.*

To imitate water samples, trip blanks are prepared in volatile organic analysis (VOA) vials with septum caps lined with polytetrafluoroethylene (PTFE). The vials are filled without headspace with analyte-free water. For soil sampled according to the requirements of EPA Method 5035, field blanks may be vials with PTFE-lined septum caps, containing aliquots of methanol or analyte-free water.

The analytical laboratory prepares the trip blanks and ships them to the project site in a cooler together with empty sample containers. The laboratory is responsible for ensuring that the trip blank and the provided empty sample containers are analyte-free. In the field, the trip blanks are kept together with sample containers and, after the samples have been collected, with the samples. *Trip blank vials are never opened in the field.*

The purpose of trip blanks is to assess the 'collected sample' representativeness by determining whether contaminants have been introduced into the samples while they were handled in the field and in transit, i.e. in coolers with ice transported from the site to the analytical laboratory. A possible mechanism of such contamination is the ability of some volatile compounds, such as methylene chloride or chlorofluor-ocarbons (Freons), to penetrate the PTFE-lined septum and dissolve in water. Potential sources of this type of contamination are either ambient volatile contaminants or the VOCs that could be emanating from the samples themselves, causing sample cross-contamination. To eliminate ambient contamination, samples must not be exposed to atmospheres containing organic vapors. Cross-contamination is best controlled by such QA measures as sample segregation and proper packaging.

Table 2.4 Field QA/QC samples and their meaning

Purpose and frequency of collection	Practical value
Trip blanks for low concentration soil and water	
Purpose: To assess the 'collected sample' representativeness Frequency: One trip blank in each cooler with samples for VOC analysis	• Useful only for water samples and soil samples with low VOC concentrations collected according to EPA Method 5035. • Enable us to determine whether ambient contamination or cross-contamination occurred during shipping and to detect contamination originating from sample containers and chemical preservatives.
Field duplicates for soil and water	
Purpose: 1. To determine total sampling and analysis precision for water samples 2. To assess inherent sample variability 3. To assess the sampling point representativeness Frequency: A common practice to collect one duplicate sample for every 10 field samples may be excessive for some projects. The DQOs should justify the need and the frequency of field duplicate collection.	• Soil duplicates allow estimating inherent sample variability. This information, if obtained during site investigation, is valuable for optimizing future sampling designs at the site. • Field duplicates increase the sampling and analysis costs by at least 10 percent. However, field duplicate precision data are rarely used for project decision-making or in evaluation of data usability.
QA samples	
Purpose: 1. To establish data comparability 2. To detect data quality problems Frequency: One for every 10 field samples	• Unless two laboratories employ the exact same procedures for extraction and analysis and samples are homogenized, results are usually comparable in qualitative terms only. • If collected and analyzed in the early stages of project work, these samples often reveal problems with sampling and analysis that can be immediately corrected. • QA samples are collected only for a small number of projects in the USA.
Temperature blanks	
Purpose: To assess the 'collected sample' representativeness Frequency Every cooler with samples	• Document that the 2 to 6°C temperature range was maintained during shipping. • Temperature blanks provide better measurements of sample temperature upon arrival to the laboratory than random measurements taken inside the cooler.

continues

Table 4   Field QA/QC samples and their meaning (continued)

Purpose and Frequency of collection	Practical value
Equipment blanks	
Purpose: To assess the collected sample 'representativeness' Frequency: Typically collected each day when non-disposable sampling equipment is decontaminated. At a minimum, one blank should be collected and analyzed when non-disposable equipment is first time decontaminated between samples to verify the effectiveness of decontamination procedure.	• Enable us to detect sample cross-contamination originating from non-disposable sampling equipment. • The only contaminants usually found in equipment blanks are common laboratory contaminants: phthalates, methylene chloride, acetone. • Occasionally, oily materials, PCBs, and VOCs retained by groundwater pumps are found in equipment blanks after samples with high contaminant concentrations.
Source water for equipment blanks	
Purpose: To verify that the water used for final rinse during decontamination is free of target analytes Frequency: Collected one time for each new batch of rinse water *only* when the quality of rinse water is questionable and when very low contaminant concentrations are a matter of concern.	• Increase the cost of analysis without enhancing data quality. • Marginally useful in *extremely rare* cases of low-level contaminant concentrations in water samples. • Contaminants found in source water are usually different from contaminats of concern.
Ambient blanks for water	
Purpose: To assess the sampling point representativeness Frequency: Every day when water samples for volatile or airborne contaminant analyses are collected.	• Enable us to determine whether airborne contaminants were introduced into water samples during sample handling in the field. • May be useful *only* when sampling water for low-level VOC or other airborne contaminant analysis is conducted in an area with high level of VOC emissions or airborne particulate matter. Due to a short exposure time to atmospheric air during sampling, these contaminants are usually undetectable in ambient blanks.

Most often, trip blank contamination originates in the laboratory, either from common airborne laboratory contaminants (methylene chloride, acetone) or from laboratory water containing VOCs, typically methylene chloride, acetone, and toluene or water disinfection byproducts (chloroform, dichlorobromomethane, chlorodibromomethane, bromoform). Rare, but well documented sources of trip blank and associated field samples contamination are insufficiently clean sample containers or contaminated preservation chemicals. The best practice to prevent this from happening is to use only pre-cleaned certified containers with preservation chemicals of documented and acceptable quality.

Aqueous trip blanks sometimes accompany soil samples collected into metal liners or glass jars. In this capacity they do not provide any meaningful information. Soil samples do not have the same contamination pathway as water samples because they are not collected in 40-milliliter (ml) VOA vials with PTFE-lined septum caps. In addition, soil does not have the same VOC transport mechanism as water does (adsorption in soil versus dissolution in water). There are other differences that do not permit this comparison: different sample handling in the laboratory; different analytical techniques used for soil and water analysis; and the differences in soil and water MDLs. That is why the comparison of low-level VOC concentrations in water to VOC concentrations in soil is never conclusive.

The use of clean oven-baked sand for soil trip blanks has also been proposed (USACE 1994; Lewis 1988). This practice does not provide any meaningful information either. The transport mechanism of VOCs in sand is different from the VOC transport mechanism in soils, which, unlike sand, have complex lithological compositions and varying organic carbon content.

Trip blanks prepared in vials and containing aliquots of methanol or analyte-free water accompany soil samples collected in a similar manner for low concentration VOC analysis according to EPA Method 5035. In this case, field samples and trip blanks have the same contamination pathway when exposed to airborne contaminants and the same VOC transport mechanism. These trip blanks provide important information, which may enable us to recognize the artifacts of improper sample handling, storage, or shipping.

Data obtained from trip blank analysis are useful in cases when low contaminant concentrations are a matter of concern. These are the concentrations within a range of 1 to 10 times of the PQL. There is always a great uncertainty associated with contaminants detected in this range as for their true origin. A comparison of the sample data to the trip blank data will determine whether contamination during sample handling and shipping has occurred or whether the sample concentrations represent the true contaminants in the sampled matrix. If samples suspected of containing high contaminant concentrations are shipped in a cooler together with low concentration samples, a trip blank will serve as an indicator of possible sample cross-contamination.

2.6.2   Field duplicates
Field duplicates are samples collected at the same time and from the same sampling point as the corresponding primary samples. As a rule, field duplicates represent 10 percent of the total number of the field samples for each matrix (one field duplicate for every 10 field samples). Field duplicate precision calculated as the RPD (Equation 1, Table 2.2) enables us to evaluate the total sampling and analysis precision for water samples and the inherent sample variability for soil samples. For soil, the extent of inherent sample variability at the sampling point is a qualitative measure of sample representativeness.

To prevent the laboratory from intentionally making field duplicate data look comparable to the primary sample data, fictitious sample numbers are assigned to field duplicates to conceal their identity. Once analytical results have been obtained from the laboratory, the project team, who has the information on the field duplicate's real identity, will compare the field duplicate and the corresponding primary sample results.

Field duplicates for water samples are collected by sequentially filling sample containers with water from a sampling device. Water sample field duplicate precision summarily measures the analytical precision and the reproducibility of the sampling procedure. Unless gross contamination in the form of a separate phase is present, field duplicate precision for properly collected and analyzed duplicate water samples is typically below 30 percent.

Duplicate soil samples may be collected as unhomogenized or homogenized split samples, or as collocated samples. Whichever sampling procedure is selected, the remaining field samples must be collected in the same manner.

Unhomogenized duplicates are soil sample aliquots sequentially collected from the same sampling point into separate sample containers. Unhomogenized duplicate sample data provide information on soil variability with respect to contaminant distribution at the sampling point. Vastly different contaminant concentrations in unhomogenized field duplicate results indicate high contaminant variability.

Homogenized splits are prepared by collecting a volume of soil from a sampling point into a bowl, thoroughly mixing it, and then placing aliquots of soil into separate containers. Homogenized field duplicate samples always have a better precision compared to unhomogenized ones and to some extent they reflect the precision of the homogenization procedures that are susceptible to human error. The process of homogenization, no matter how thoroughly performed, sometimes cannot eliminate the effects of matrix variability.

Homogenization can be conducted for non-volatile compounds only, by mixing and splitting soil according to defined procedures. Field duplicates for VOC analysis cannot be homogenized due to analyte volatilization during mixing. These duplicates must be collected either as unhomogenized or as collocated samples.

Collocated duplicates are collected into two separate containers from two sampling points located in the immediate proximity of each other. Normally, collocated duplicates are not homogenized. The collocated duplicate precision reflects only the matrix variability with respect to contaminant distribution.

In each of these soil field duplicate collection techniques, matrix variability is a decisive factor that cannot be entirely controlled. Consequently, field duplicate RPD cannot be controlled and it should not be compared to a numeric standard, such as an acceptance criterion. Soil field duplicates are best assessed qualitatively by drawing conclusions from the comparison of the identified contaminants and their concentrations.

Substantial sample variability affecting field duplicate precision and, consequently, the total sampling and analysis precision is to be expected for many types of soil samples and even some types of water samples as shown in Example 2.5. High sample variability has been documented for any sample collection technique for VOC analysis (Vitale, 1999; Popek, 1999). Depending on the project DQO, the collection of field duplicates for samples with expected high variability may be not necessary.

Example 2.5:   Samples with expected low field duplicate precision

Low field duplicate precision may be expected for the following types of samples:

- Soil samples for VOC analysis
- Unhomogenized, homogenized, and collocated soil samples collected for waste oil, other heavy petroleum fuels, or PCB analysis
- Unhomogenized, homogenized, and collocated soil samples collected from shooting ranges for metal analysis
- Soil samples with high clay contents: they cannot be effectively homogenized and have a high variability of contaminant distribution
- Soil samples with high gravel content
- Water samples with floating petroleum product—these samples should not be collected altogether
- Groundwater samples collected with direct push sampling tools
- Multiphase waste samples
- Soil vapor samples

Contaminant distribution in soil and water depends on such factors as soil properties; the physical and chemical properties of the contaminant; contaminant fate and transport in soil, groundwater or surface water; and even the manner in which the contaminant was introduced into the environment. The knowledge of these issues coupled with available information on site history and background allows us to make valid assumptions in the planning phase on contaminant distribution and variability at the site.

It is particularly important to consider all these factors while developing soil sampling designs. The frequency of field duplicate sample collection should be based on the usefulness of the information they will provide. If contaminant distribution is expected to be sporadic, to better characterize the site we should collect additional field samples instead of duplicate samples or consider compositing as part of the sampling design. If no information on variability is available, a few duplicates collected early during site investigation will provide the data for making educated decisions related to future sampling at the site.

2.6.3   Equipment blanks

Equipment blank is a sample of water collected from the surface of a decontaminated sampling tool to verify the effectiveness of a cleaning procedure. Equipment blanks are sometimes called ***rinsate blanks***. They are collected as samples of the final rinse water from non-disposable sampling tools after they have been cleaned between samples. The field crew pours analyte-free water over the tool's surface that has come in contact with the sampled medium. The water is diverted directly into sample containers and analyzed for the project contaminants of concern.

If analyte-free water is not available, the field crew may use tap or bottled water for the final rinsing. A sample of source water, called the *source blank* may be also analyzed for the project contaminants of concern.

Equipment blanks enable us to assess the 'collected sample' representativeness. The purpose of collecting equipment blanks is to detect the presence of contamination from the sampling equipment itself or any cross-contamination with previously collected samples. For example, metal liners for core barrel or split spoon samplers are not always precleaned by the manufacturer or distributor. They must be cleaned in the field prior to sampling to eliminate the potential for sample contamination.

Non-dedicated pumps used for well purging and sampling are another example of sampling equipment that must be decontaminated between samples. Contaminants from the water that comes in contact with the pump and the plastic tubing are adsorbed onto their surfaces. If the pump and the tubing are not properly decontaminated, the adsorbed contaminants will be gradually released as the water from a different well travels through the pump system. This is particularly true for water with high VOC concentrations. Generally, flexible polymeric materials have a greater capacity for sorption than the rigid ones; the greater the flexibility, the stronger is the retention (Parker, 1997). Non-permeable materials (glass, metals) do not have this property and can be easily cleaned of surface contamination.

The intent of equipment rinsate blank collection as a field QC sample seems reasonable. In reality, however, equipment blank analyses rarely provide information that can be meaningfully related to the field samples because the only contaminants that are usually present in equipment blanks are common laboratory contaminants or byproducts of water disinfection process.

Generally, equipment blanks provide useful information *only* if all of the following conditions are met:

• The sampled medium is water.
• Low concentrations of organic contaminants and metals are a matter of concern.
• Non-disposable or disposable but *not precleaned* sampling equipment is used.

Existing EPA and DOD guidance documents specify a universal frequency of equipment blank sample collection as one per day or one per sampling event, whichever comes first, for all projects where sampling equipment is cleaned. The prescriptive collection and analysis of equipment blanks increases the total project cost and, considering the low value of obtained information, seems excessive. A widely used practice of equipment blank collection for soil sampling equipment is particularly irrational, as soil and water concentrations cannot be compared in a meaningful manner. Collecting equipment blanks for VOC analysis from soil sampling tools made of impervious materials defies the physical property of volatility itself.

Equipment blanks are relics of the non-disposable sampling equipment era. Today we use a variety of disposable equipment to collect a vast majority of environmental samples. Precleaned disposable sampling equipment is available for most types of sampling, and the wide selection of modern non-adsorbing materials available today enables us to reduce or even eliminate the risk of cross-contamination.

What is the criterion that determines whether a decontamination procedure was effective? It is not the absolute absence of any chemical contaminants in the equipment blank. Important for the project are only the contaminants of concern and their concentrations. For site investigations, when no information is available on existing pollutants, it may be important that no contaminants of concern are present in equipment blank samples above the laboratory PQLs. On the other hand, for site remediation projects, the presence of contaminants of concern in equipment blank samples may be acceptable, if these concentrations are only a fraction of the action levels. ***The decision to decontaminate equipment and the selection of the acceptability criteria for equipment blanks are made in the DQO process based on the intended use of the data.***

In the planning phase we should consider the overall cost of equipment decontamination, which may be summarized as follows:

- Decontamination procedures consume a great deal of the sampling crew's time.
- The collection of equipment blank samples increases the total cost of sample handling and shipment.
- Equipment blanks create additional analytical and data management costs.
- The storage, analysis, handling, and disposal of large volumes of decontamination wastewater increase the total project cost.

Instead of applying the DQO-blind protocol requirements, we should always determine the need and extent of decontamination based on the intended use of the data. Understanding the DQOs will allow the project team to develop decontamination procedures that are cost-effective and appropriate for the contaminant concentration range being measured and the action levels applied. The following questions need to be answered in the planning phase:

1. What is the proper decontamination procedure?
2. What data are we going to obtain from the analysis of equipment blanks?
3. Will equipment blank data affect our decisions?
4. How many equipment blank samples are sufficient for documenting the cleanliness of sampling equipment?

When selecting a decontamination procedure, we should choose the most appropriate one for the matrix and the contaminant. Soil sampling equipment decontamination procedures are basically uniform, whereas there are a variety of different procedures for decontaminating of water sampling equipment (Parker, 1995). When sampling for water, we should carefully select the types and materials of the pumps, the tubing, and the bailers, and consider the order in which wells with different contaminant concentrations will be sampled. To avoid or reduce equipment decontamination, we may plan on using dedicated sampling pumps, dedicated tubing, and disposable bailers for water sampling and disposable plastic scoops for soil sampling.

The collection of equipment blank data is unavoidable for the types of equipment that require decontamination prior to each use (for example, non-disposable bailers, non-dedicated submersible pumps, augers, split spoons). However, the number of equipment blank samples may be reduced using a strategy demonstrated in Example 2.6.

Example 2.6:   Equipment decontamination strategy

One equipment blank collected after the sampling equipment has been cleaned the first time may often provide a satisfactory resolution of this issue. Once a decontamination procedure has been established and verified by an equipment blank analysis, the application of the same procedure should render the same level of cleanliness of the sampling equipment. If site conditions suddenly change with respect to the type of the sampled medium or the contaminant nature and suspected concentrations, another equipment blank sample may be collected to verify that the change in conditions did not reduce the effectiveness of the established decontamination procedure.

2.6.4   Temperature blanks

To reduce the loss of contaminants from the collected samples due to volatilization and bacterial degradation, the majority of environmental samples must be preserved by storage within a temperature range of 2–6 °Celsius (C). Samples collected for VOC analysis are particularly susceptible to such losses even at slightly elevated temperatures. Preserved samples represent the sampling point conditions at the time of sampling, and that is why proper preservation upon sample collection is necessary to assure the 'collected sample' representativeness.

A temperature blank is a container with tap water, which is shipped together with samples from the field to the laboratory. To create cold storage conditions, the samples and the temperature blank are shipped on ice in insulated coolers. The purpose of the temperature blank is to establish whether the temperature range of 2–6 °C has been exceeded in the cooler while in transit. The laboratory personnel will measure the temperature of the water in the temperature blank container and thus will document this aspect of sample preservation.

If a temperature blank is not enclosed in the cooler with samples, the laboratory personnel will randomly measure the temperature in any part of the cooler, and this temperature may or may not accurately represent the true condition. Temperature blanks, being simulated samples, provide an accurate measurement of field sample temperature upon arrival to the laboratory.

2.6.5   Ambient blanks

Ambient (field blanks) are sample containers with PTFE-lined septum caps filled with analyte-free water in the field to establish whether contamination could have been introduced into water samples from ambient air during sampling. The laboratory provides a bottle of analyte-free water, and the field crew pours this water from the bottle into a sample container in a manner that simulates the transfer of a sample from a sampling tool into a container. Ambient blanks are analyzed for the contaminants of concern that may be airborne at the site in order to assess the sampling point representativeness.

Ambient blanks are intended to detect airborne contamination that may be affecting samples collected in atmospheres with high contents of organic vapor. This type of ambient contamination may be present at airport runways, refineries, gasoline

of sample types, sampling points, and sampling frequencies. Maps showing sampling locations, such as grid designs, positions of soil borings and groundwater monitoring wells, are always included in the QAPP. The types and quantities field QC and QA samples; the sampling and analytical methods requirements; the sampling procedures; the requirements for sample containers and custody, sample packaging and shipping to the laboratory and for the decontamination of sampling equipment are also described. Relevant SOPs may be included into the QAPP as appendices.

This group of elements contains a large volume of information on analytical laboratory method requirements and procedures. Therefore, the laboratory that will conduct the analysis should provide this information to the project team for the incorporation into the QAPP. For example, the selection of analytical laboratory methods and QC requirements definitely needs input from the analytical laboratory, particularly if low detection limits or non-routine analyses are concerned. Analytical instrument calibration and maintenance requirements should also be developed as a cooperative effort with the analytical laboratory or by the individuals who are well-versed in laboratory practices and procedures.

Procedures for the calibration of field instruments and inspection of field supplies to determine whether they meet the project needs are important factors in the implementation of field activities. Element B9, cryptically named by the EPA 'Non-Direct Measurements' is applicable when previously collected data are to be used for project decisions. Data management, which consists of field and laboratory data recording, transmittal, tracking, storage, and retrieval, is described in element B10.

Group C. Assessment/Oversight
    C1   Assessments and response actions
    C2   Reports to management

The assessment of project activities may include performance and systems audits, data quality audits, peer review, PE samples, as appropriate for a given project. Assessment enables project personnel to identify field and laboratory problems or variances from the project scope and to implement timely corrective action. The findings and response actions originating from assessment activities are documented in reports to management.

Group D. Data Validation and Usability
    D1   Data review, verification, and validation
    D2   Validation and verification methods
    D3   Reconciliation with user requirements

Group D elements describe the procedures that will be used for the assessment of data quality and usability. Properly conducted laboratory data review, verification, and data validation establishes whether the obtained data are of the right type, quality, and quantity to support their intended use.

The QAPP may be a comprehensive document that addresses a complex, multiple-task project to be executed over the course of several years, or it may address a single, limited task within a larger project. A comprehensive QAPP that addresses long-term activities will require revisions and updates as new analytical methods and sampling requirements are introduced. The greatest drawback of a long-term QAPP is its lack of flexibility. Once prepared for a project with certain DQOs, a comprehensive QAPP becomes a prescriptive document and is often inapplicable for the new tasks with different DQOs.

2.8.2 Sampling and Analysis Plan format

The information contained in Groups A, B, C, and D can be logically broken down into two distinctive parts that describe the field and analytical laboratory objectives, with each part presented in a separate document. The elements that address field activities are removed from the QAPP and segregated into a document that is called Field Sampling Plan (FSP). The two documents, the FSP and the QAPP, constitute the Sampling and Analysis Plan. A short introduction to the SAP may describe the purpose of its two parts and tie them together under a single cover.

The advantage of the SAP format is a clear division of the field and laboratory activities and the ensuing focus of the field and laboratory personnel on their respective tasks. The USACE SAP format by being extremely detailed serves as a comprehensive model for the preparation of the sampling and analysis plans (USACE, 1994). The disadvantage of the USACE format is a certain redundancy, as the same information on project description, history, background, organization, and schedule needs to be repeated twice, in the FSP and in the QAPP. (In practice, once these sections have been included in the FSP, they may be incorporated into the QAPP by reference.)

The preparation of a planning document, whether it is called the QAPP or the SAP, can be a complex, time-consuming, and expensive task. (For the sake of brevity and clarity, in this Guide we refer to this planning document as the SAP.) Much of the information presented in the SAP originates from the analytical laboratory and is available to the planning team from the Laboratory QA Manual. Extensive planning documents are usually prepared from boilerplate electronic texts, and original writing may constitute about one third of the whole document.

However, not every environmental project requires a comprehensive and elaborate planning document. Usually the contents and the format of the SAP are determined by such factors as contractual requirements, regulatory agency oversight, or by financial constraints. A smaller document that contains the most essential SAP elements in combination with the Laboratory QA Manual may be as functional as a full-scale SAP. Appendix 7 presents an example of a SAP table of contents. A SAP prepared in this format is a very effective document that communicates the information essential for project implementation and assessment phases without being overloaded with information available from other sources.

To assist the reader in the planning document preparation, whether it is a QAPP or a SAP, Table 2.5 cross-references the required EPA QAPP elements with the subject matter chapters of the Guide.

Table 2.5   EPA QAPP elements in the Guide

QAPP element	Corresponding Guide chapter
A4 Project/task organization A5 Problem definition/background A6 Project/task description	2.1 What are the Data Quality Objectives? 2.2 Definitive, screening, and effective data 2.3 Regulatory overview and action levels
A7 Quality objectives and criteria for measurement data	2.1.9 Data quality indicators
A9 Documents and records	3.10 Field records 4.3.7 Data package preparation
B1 Sampling process design	2.5 Sampling strategies
B2 Sampling methods	3.5 Soil sampling designs 3.6 Water sampling 3.7 Surface sampling with wipes
B3 Sample handling and custody	3.2 Sample custody and tracking 3.3 Sample preservation techniques 3.4 Sample packaging and shipment 3.8 Equipment decontamination
B4 Analytical methods	2.4 How to navigate the analytical method maze 2.9 Analytical method and laboratory selection 4.4 Analytical techniques and their applications
B5 Quality control	2.6 Field quality control and quality assurance samples 4.6.1 Laboratory quality control samples and their meaning
B6 Instrument/equipment, inspection and maintenance B8 Inspection/acceptance of supplies and consumables	2.10 Preparedness and coordination 3.3 Sample preservation techniques 3.9 Field measurements
B7 Instrument/equipment calibration and frequency	3.9 Field measurements 4.5.2 Importance of calibration
B10 Data management	4.3.7 Data package preparation
C1 Assessment and response actions C2 Reports to management	4.6.3 Systems and performance audits
D1 Data review, verification, and validation D2 Validation and verification methods	4.3.5 Data reduction, verification, and reporting 4.3.6 Internal data review 5.1 Data evaluation 5.2 The seven steps of data evaluation
D3 Reconciliation with user requirements	5.3 The seven steps of the data quality assessment

2.8.3 Laboratory Statement of Work

The analytical laboratory is an integral member of the project team, and it should have a full access to the planning documents that describe the requirements for environmental data collection. That is why it is always a good practice to provide the SAP to the laboratories, which will be conducting the project analytical work.

Another important planning document is the Laboratory Statement of Work, which is based on the SAP requirements for laboratory analysis. The Laboratory SOW summarizes project analytical requirements and serves as a technical specification during the laboratory selection process. The project team sends the SOW to several qualified laboratories for the preparation of a price proposal. At a minimum, the SOW should include the following items:

- A brief project description
- Requirements for laboratory accreditation
- Number of samples for each matrix, including field QC samples
- Types of analysis and appropriate analytical methods
- Target analytes and reporting limits (RLs) as required by the project DQOs (RLs are defined in Chapter 4.5.1.)
- Accuracy and precision acceptability criteria
- Requirements for analytical instrument calibration and laboratory QC checks
- Data package content requirements
- Format for electronic data deliverables (EDDs)
- Sampling schedule
- Turnaround time for laboratory analysis and data package delivery
- The name and address of the project point of contact

In order to minimize a risk that the project DQOs will not be met due to insufficient quality of data or wrong types of analysis, we must clearly convey all of these requirements to the laboratory prior to start of work. A well-prepared laboratory SOW serves not only as a procurement tool but also as a means to ensure that the project schedule and the DQOs are clearly communicated to a prospective team member, the analytical laboratory.

Laboratories review the SOW to establish whether their capacity, capability, and accreditations meet the project requirements. If any of the requirements are not met, the laboratory will specify the exceptions in a letter accompanying the bid proposal. Typically, exceptions include the absence of necessary accreditations and variances from the desired RLs and acceptability criteria. If the planning team has selected the RLs and the acceptability criteria based on current analytical methods and standard laboratory practices, the variances will not be significant and will not affect data quality or usability.

2.9 ANALYTICAL METHOD AND LABORATORY SELECTION

On every project we face decisions related to the selection of proper analytical methods and a laboratory to perform them. Various regulatory protocols offer hundreds of analytical methods. How do we identify the ones that are appropriate for the project? There are dozens of analytical laboratories competing for work. How do we find the one that will produce the data of a required quality at a price that the project can afford? These are difficult questions. And we find the answers to them in the DQO process. Applying a step-wise approach and relying on decisions formulated during the DQO process will help us arrive at a correct choice.

The selection of proper analytical methods and the best laboratory to perform them may be described as a seven-step process shown in Figure 2.6.

The groundwork for laboratory selection starts during the planning phase of the data collection process. By having identified the intended use of the data (Step 1), the regulatory framework (Step 2), and action levels and appropriate reporting limits (Step 3) in the course of the DQO process, we narrow down the alternatives for the selection of analytical methods. By evaluating the possible matrix limitations (Step 4), such as matrix interferences and their effects on the RLs, we will further refine the

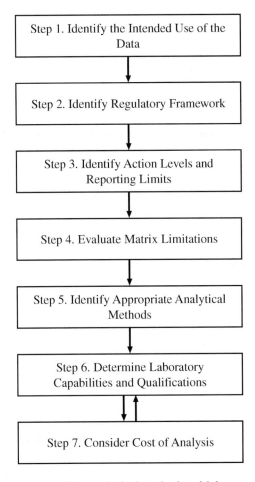

Figure 2.6   The seven steps of the analytical method and laboratory selection process.

selection and will finally identify the methods that allow us to meet the project objectives (Step 5).

We will document the identified analytical methods, numbers and types of samples, and appropriate QC requirements in the SAP, prepare the Laboratory SOW, and send it to at least two laboratories for competitive bidding. To be thorough in the selection process, we should request that the laboratories supply with their price proposal a description of their technical capabilities and sample throughput capacity and the QA Manual that documents the laboratory QA system. This information may not be necessary if a laboratory has been providing analytical services to us in the past, but it will be a prudent practice if new laboratories are among the bidders.

Most environmental laboratories have a Statement of Qualifications (SOQ) document that establishes the laboratory's qualifications through descriptions of project experience, a list of accreditations, and resumes of laboratory staff. A review of the SOQ together with the Laboratory QA Manual will enable us to establish the technical capabilities and the capacity of each laboratory (Step 6). Applicable accreditations should be verified as part of laboratory qualifications, such as the state and DOD certifications or approvals.

Only after the qualified laboratories have been identified should we evaluate the proposed prices (Step 7). At this step, we will be able to select the lowest bidder laboratory, which is qualified to meet the technical demands of the project and is capable of meeting the project schedule for sampling and analysis. The schedules, however, are subject to frequent changes, and delays in sampling and sample delivery to the laboratory may force the laboratory to fail on their commitment to the project. In this case, the next lowest bidder backup laboratory, which has been already evaluated in the selection process, may be able to accept the work.

Following are three general rules for the analytical method selection process:

Rule 1. Intended use of the data and the regulatory framework guide analytical method selection.

- Water discharges to the ground (into an infiltration trench or a storm drain) or a body of water are regulated by the CWA and usually require an NPDES permit. Analytical methods and discharge limitation concentrations (action levels) will be specified in the permit.
- Waste disposal and treatment facilities have their own requirements for acceptance of waste. For example, wastewater may be discharged into sanitary sewer if it meets the requirements of the POTW batch discharge permit. This permit will list the analytes, methods, and discharge limitation concentrations (action levels). Disposal facilities (e.g. landfills) have lists of analyses that must be performed as a condition of waste acceptance by the facility.
- Soil and groundwater samples from Superfund sites require CLP analysis for the TCL and TAL parameters; other analytical methods, such as the SW-846 methods, may be used for the parameters that are not part of the TCL or the TAL.
- Soil and groundwater from non-Superfund sites are typically analyzed by SW-846 methods.

- Waste characterization analyses are conducted according to SW-846 methods.
- Groundwater monitoring may be conducted by SW-846 methods or by the methods for drinking water analysis.
- Petroleum fuel analyses in some states are conducted according to state-specific methods, whereas in others they are conducted according to EPA Method 8015.

Rule 2. The RL selection is guided by the intended use of the data, such as a comparison to the action levels or discharge permit limitations; risk assessment; waste disposal.

- The RLs must be lower than the action levels.
- Generally, laboratory MDLs must be below the RLs. One of the rare exceptions from this rule may be reporting of vinyl chloride in water at the MDL for some data uses.
- Different analytical techniques have different detection capabilities for the same analytes. We should choose methods with the PQLs that are at least two times lower than the action levels.
- Risk-based action levels are calculated concentration values that may be lower than the current detection capabilities of analytical methods. In such cases, a PBMS, specifically developed to meet an unusually low detection limit, may be a solution. However, some calculated risk-based action levels cannot be attained even with the most advanced analytical techniques.
- If extraordinary low action levels cannot be attained by the BAT analytical method, the MDLs for this method may become the action levels.

Rule 3. Matrix interferences may have a detrimental effect on analytical results and reporting limits.

- Matrix interferences that cannot be overcome even with most rigorous cleanup procedures are a bane of environmental sample analysis because they cause elevated RLs.
- Oily matrices have a negative effect on the RLs of individual analytes. These effects cannot be reduced by the application of cleanup techniques, in particular when target analytes are part of the oil composition.
- Interferences from naturally occurring organic compounds are a common source of elevated RLs and false positive results in petroleum fuel analysis by EPA Method 8015, pesticide analysis by EPA Method 8081, and herbicide analysis by EPA Method 8151.
- High concentrations of metals in some types of soils or industrial sludges may cause false positive results and elevated RLs for other metal constituents.
- Turbid groundwater with poor clarity often produces elevated RLs for hexavalent chromium analysis by the colorimetry technique.

There are three general rules for the laboratory selection process:

Rule 1. Laboratory selection should never be based on price.

- Selecting a laboratory based solely on the low price of analysis is a regrettable practice. While selecting a laboratory, we should primarily focus on the laboratory's ability to provide data of known and documented quality. The issue of price should be considered *after* several laboratories with equal technical capabilities have been identified; only at this point the lowest price qualified laboratory may be selected.

Rule 2. A state accreditation is not a guarantee of acceptable data quality.

- A common misconception is that all environmental laboratories provide the same level of quality because they are accredited by a state agency. Laboratory audits and accreditations do not produce valid data; solid technical practices do.

Practical Tips: Analytical method and laboratory selection

1. Verify laboratory accreditations: some states grant a blanket approval for all analytical methods based on one series of methods, the others approve different series and different methods within the series separately.

2. Laboratories accredited by the DOD branches usually have better quality systems than the laboratories that work for private industry clients only.

3. Ask the laboratory to provide client references; call the references for information on laboratory performance.

4. As part of prequalification process request the recent results for the EPA PE sample studies to evaluate the laboratory's ability to correctly identify and accurately quantify target analytes.

5. Verify financial stability of the laboratory; laboratories have been known to go out of business without notifying the client, thus jeopardizing the course of project activities.

6. Verify that the laboratory has an effective ethics and data integrity program and that each hired person is required to sign an ethics agreement.

7. Verify whether the laboratory is not on the *List of Parties Excluded from Federal Procurement and Nonprocurement Programs*, particularly when selecting laboratories for government project work. The list can be found on the Internet at http://www.arnet.gov/epls.

8. Inquire whether the laboratory screens the new hires to make sure that debarred professionals are not assigned to work on government projects.

9. Whenever possible, select a qualified laboratory that is located close to the project site or home office. This will reduce expenses for a laboratory audit, if such is planned or if a need to conduct one arises.

10. Pay an informal visit to the laboratory to evaluate general housekeeping practices: a cluttered, untidy, or dirty laboratory is not a good sign.

Rule 3. Good laboratory service is not necessarily an indicator of good laboratory analysis.

- Environmental professionals often mistake good laboratory service for good laboratory analysis. It is true that a responsive, client-oriented laboratory reaches out to its clients to assist them in any way and makes a good impression. But good laboratory service is not equal to good analysis, and we should be able to tell the two apart.

2.10 PREPAREDNESS AND COORDINATION

Another important aspect of the planning phase is preparation for sampling, which may include some or all of the following actions:

- Identify and secure a subcontractor for specialized sampling, such as the drilling for the installation of monitoring wells and placing of boreholes for soil sampling
- Identify members of the sampling crew, who are qualified to perform the required sampling tasks
- Identify and obtain the necessary field sampling equipment and supplies
- Provide a copy of the SAP to the selected analytical laboratory
- Obtain sample containers and shipping supplies
- Obtain calibrated field instruments and screening test kits
- Conduct a premobilization ('kick-off') meeting or a conference call with all parties involved to discuss the upcoming sampling events

These *premobilization activities* take place not long before the project is mobilized to the field. Premobilization activities require a great deal of organizational skills and, depending on the project scope, may take several days. The most important issues in premobilization planning are communications with the field crews and, thoughtful, systematic, efficient preparation.

The analytical laboratory is an important but often overlooked member of the project team; good communications with the laboratory are essential for the project success. The laboratory that conducts analysis of project samples will usually provide the sample containers, coolers for shipping samples, and the packing materials. The laboratory should have as much advanced notification as possible to prepare the appropriate containers and ship them to the sampling crew. After the sampling containers have been received, a member of a sampling crew should verify that the quantities and preservation chemicals are correct and meet the SAP requirements.

The laboratory must also know the project schedule for internal planning of incoming samples. It is particularly important when expedited turnaround time of analysis is required. Any last minute change in the sampling and analysis design or in the project schedule should be always communicated to the laboratory.

If the laboratory does not provide the sampling containers, a member of the project team should identify the vendors and purchase the containers that meet the SAP requirements. Obtaining proper field sampling equipment and supplies is usually a responsibility of the sampling crew. This may constitute an extensive list ranging from bailers and hand augers to such mundane but vital items as pens and shipping tape.

A list of field supplies may be quickly put together by reviewing the SAP and the procedures for the sampling to be performed. Using a sampling checklist, similar to one shown in Appendix 8, will facilitate efficient preparation.

The sampling crew is also responsible for obtaining field instruments and test kits, if required by the project scope. Field instruments should be calibrated prior to mobilization; as a rule, they come with manufacturer's instruction manuals and calibration standards. Field test kits, which became widely used in the recent years, are also obtained in advance in sufficient quantity and of the quality specified in the SAP. Because different manufacturers make field instruments and test kits in different formats, members of the sampling crew may require special training in their use.

Practical Tips: Preparedness and coordination

1. Inform the laboratory of upcoming sampling events and changes in the sampling schedule.

2. Send the SAP to the laboratory prior to start of sampling activities.

3. Obtain the sample containers and verify their appropriateness for the samples to be collected.

4. Make a checklist of field sampling equipment and supplies.

5. To identify the necessary tools and supplies, read the sampling SOPs and mentally visualize the steps of the sampling process.

6. Carry a toolbox or a sturdy plastic container with general supplies that are always needed in the field (field record forms, writing paper, pens, shipping tape, a tape measure, a compass, etc.).

7. Inform the laboratory every time samples are being shipped.

8. In the planning phase, identify a courier service for sample delivery to the laboratory.

9. Obtain and calibrate field instruments, make sure that the manuals and calibration standards are available for future use in the field.

10. Obtain field screening kits and all necessary supplies and verify that the quantity of supplies is sufficient for the project task.

11. Always obtain spare field screening kits for backup in case more samples are collected than planned.

12. Make sure that field personnel are trained in the use of these particular models of field instruments and test kits.

13. Conduct a project kick-off meeting with the laboratory and project staff to clarify the sampling schedule and analytical requirements.

3

Practical approach to sampling

After the planning phase of the data collection process has been completed and the foundation of the data collection pyramid has been built, the project moves into its second phase, implementation. The implementation phase takes place in the field and at the laboratory. This chapter addresses the tasks of field implementation, such as Task 3—Sampling and Field Data Collection and Task 5—Field QA/QC, shown within the data collection pyramid in Figure 3.1. The main features of these field tasks are sampling procedures; sample custody and tracking; preservation techniques; equipment decontamination; field screening; and record keeping.

Project implementation cannot be carried out without planning and it can be only as good as the planning itself. Only thorough and systematic planning provides a firm foundation for successful implementation. Poor planning typically leads to ineffectual implementation, which in turn does not produce the data of the type, quality and quantity required for reliable decision-making.

In environmental chemical data collection, the purpose of field implementation is to collect *representative* samples for the production of *valid and relevant data*. To be representative, the samples must have all of the following attributes:

- Be collected from the sampling points selected in the course of a systematic planning process
- Be collected according to appropriate sampling procedures
- Be unaffected by ambient contamination and cross-contamination from other samples
- Be placed in proper sampling containers and correctly preserved
- Be accompanied by field QC samples justified by the project DQOs
- Have a traceable custody chain and reliable field records

The requirements for the collection of representative samples are documented in the SAP. If all of these requirements are met in the course of field implementation, the collected samples will be representative of the population of interest. The goal for sampling completeness will be also met. Deviations from the SAP requirements have a potential to degrade the DQIs of representativeness and completeness and, consequently, the usability of a data set. The most damaging errors that may take place during field implementation are the collection of unrepresentative samples; incomplete sampling that produces data gaps; and incomplete documentation that

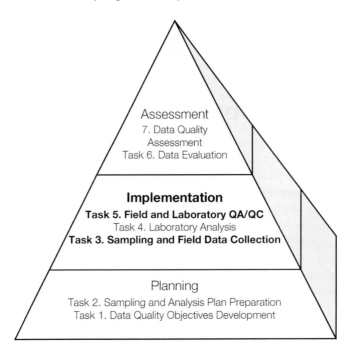

Figure 3.1   Field implementation phase of environmental chemical data collection.

does not allow the reconstruction of the sampling effort or makes the collected samples legally vulnerable.

Sampling of surface and subsurface soil, stockpiled soil, groundwater and surface water, and sampling with wipes constitute the majority of sampling applications conducted for environmental pollutant data collection. This chapter provides step-by-step procedures for sampling of these common environmental matrices for a wide range of chemical properties and chemical contaminants. The use of these procedures as baseline information will enable project teams to create general and project-specific SOPs for sampling and sample handling in the field.

3.1   SEVEN STEPS OF THE SAMPLING PROCESS
The sample collection process may be broken down into seven consecutive steps shown in Figure 3.2. Each step, if not completely understood and properly carried out, can potentially invalidate the collected samples and even nullify the whole sampling effort.

The sampling process starts with the reading and understanding of the SAP, which addresses the project DQOs and describes the specific procedures to be used during sampling (Step 1). A well-written SAP will provide the field sampling team with sufficient detail for conducting proper sampling, sample handling, and for completing field documentation.

Based on the information gained from the SAP, we will identify the types and quantities of the needed sampling equipment and supplies (Step 2). A thoughtfully

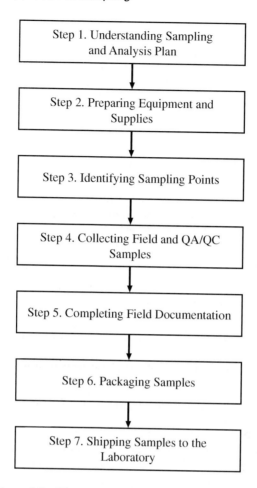

Figure 3.2   The seven steps of the sampling process.

prepared SAP will include a list of sampling supplies and equipment to facilitate the efficient preparation for sampling.

Once in the field, we will refer to the SAP for finding the sampling points (Step 3). The SAP will contain maps, sampling grids, treatment system process diagrams, etc., to direct us to the correct sampling points. Only after the sampling points have been identified, we can start the collection of field and QA/QC samples according to the procedures described in the SAP (Step 4).

Completing field documentation (Step 5), such as the *Chain-of-Custody (COC)* Form, field logs, and sampling forms, is a separate and distinctive step in the sampling process. Field documentation establishes the basis for informed data interpretation and efficient and accurate report preparation. The COC form is usually the only written means of communications with the analytical laboratory. It also serves a legal function by documenting the chain of individuals, who were responsible for sample integrity.

Packaging the collected samples for shipment (Step 6), a tedious task as it may be, requires a certain level of experience. Improperly packaged samples are often broken in transit; insufficient quantities of ice do not provide the proper temperature control—all of these events have a potential to invalidate the collected samples and, in fact, are a common cause for resampling.

After the samples have been packaged, they are shipped to the laboratory (Step 7). Samples must reach the laboratory in the shortest period of time possible, as any delays in transit jeopardize temperature conditions inside the shipping container and erode sample holding time. Incorrect information on shipping labels and lack of coordination with the delivery service are known to cause delays of sample delivery to the laboratory.

Practical Tips: **Seven steps of the** *sampling process*

1. Have the SAP on hand while in the field and refer to it frequently for sampling guidance.

2. If the SAP does not have sufficient detail on the type of sample containers, contact the laboratory for help.

3. Have on hand the telephone numbers of useful contacts at the home office and at the laboratory. When in doubt, call another project team member for help and advice.

4. Obtain the sampling equipment and field instruments ahead of time and verify their working condition.

5. Complete field documentation on the day of sampling and while still in the field; do not delay till the next day.

6. Have ample packing material and bagged ice on hand and give yourself sufficient time to pack the samples into shipping containers.

7. Verify the shipping address and telephone number of the laboratory ahead of time.

8. Identify the courier or an overnight delivery service prior to mobilizing to the field.

9. When collecting samples that are of critical importance or that cannot be resampled, always consider the collection of backup samples.

10. Plan ahead, as each task usually takes longer than expected.

11. When in doubt, do not assume anything; verify fact before taking action.

3.2 SAMPLE CUSTODY AND TRACKING

The collected samples must be identified with unique sample numbers; efficiently tracked in the field; stored in a secure location for the preservation of sample integrity; and transferred to the laboratory with the information on their identification (ID) and the requested analysis. The provisions for sample numbering, labelling, storage in the field; tracking and transfer to the laboratory are set forth in the SAP.

3.2.1 Chain-of-Custody Form

Environmental samples may occasionally become legal evidence, and their possession must be traceable. An official document called the *Chain-of-Custody Form* serves the purpose of documenting sample transfer from one party to another. It is a record that tracks samples through their complete life span and serves as analysis request form. Over the years, the use of the COC became the industry standard regardless of whether the samples are being collected for legal purposes or not.

A sample is under custody, if one or more of the following criteria are met (DOE, 1996):

- It is in the sampler's possession
- It is in the sampler's view after being in possession
- It is in a designated secure area

Each analytical laboratory has its own COC form created based on the laboratory's own experience and operating practices. All of these different COC forms must, however, have the following information in common:

- Project name and point of contact
- Signature of a sampling team member
- Field sample ID number
- Date and time of sampling
- Grab or composite sample designation
- Signatures of individuals involved in sample transfer
- Air bill or other shipping number, if applicable
- Analytical requirements

Additional information usually includes such items as the description of sample matrices and sampling points; the number of containers for each sample; preservation chemicals; and special requirements or instructions. Appendix 9 shows an example of a COC form.

COC forms are usually printed as white originals with one or two attached carbon reproduction copies of different colors. A sequential number often (but not always) identifies each form; this number is useful for sample tracking. The carbon copy stays at the project site as part of project records, and the original is sent with samples to the laboratory. The laboratory will in turn include a fully completed original COC form with the sample data package. (Technically, a fully completed COC form should include a record of sample disposal, but in reality this never happens because laboratories document sample disposal with internal records).

There is no EPA requirement for the COC form to be produced with carbon copies. These are useful only if there is no copy machine at the project site. If a copy machine is available, photocopying the original COC form for project record keeping is perfectly acceptable. Generating a COC form with a computer program in the field or prior to start of field work is the best way to create a legible and legal document. Computer programs that exist today allow printing sample labels and COC forms, tracking samples electronically, and exchanging information with databases.

Field sampling personnel will initiate a COC form as the samples are being collected. After samples have been collected and packed for shipment, a sampling team member will sign the COC form and record in it the air bill or way bill number of a commercial delivery service like Federal Express or United Parcel Service. The COC form is then placed in a plastic bag and taped to the inside of the cooler lid. Representatives of the shipping companies are not required to sign the COC forms. If samples are relinquished to a laboratory courier, the courier signs the COC form as a receiving party.

Documenting sample transfer from one person to another is a critical legal issue, and many legal claims have been dismissed in court based on incomplete or incorrect COC forms. The sampling team must fill out the COC form with great care, not only because it may become a legal document but also because it is often the *only* written record related to sample identification, type, and requested analysis that the laboratory will get.

Another part of legal documentation used in sample handling is the custody seal shown in Appendix 10. A custody seal is a narrow paper or plastic strip with adhesive backside. Custody seals were introduced for use on sample containers with samples explicitly collected for projects in litigation. A signed and dated custody seal placed in the field on each sample container connects the container and its lid. A broken custody seal discovered by the laboratory in the course of sample receiving indicates that a sample has been tampered with. There is no reason to use custody seals in this manner on projects that are not under litigation. However, as an external security measure, we may place two custody seals on the front and rear sides of the shipping container lid.

3.2.2 Sample numbering and labelling

Each collected sample must have its own unique identification number. The project staff will use this number to track the sample in the field and at the home office during the project assessment phase. The laboratory will use this number for matching the sample and the requested analysis. The laboratory will enter the field sample IDs into the data base and assign the samples with laboratory IDs for internal tracking. Analytical results will be reported for the samples identified according to their field and laboratory IDs.

In the course of sample tracking, data evaluation, and interpretation, field sample IDs may be entered into several different field forms, spreadsheets, and data bases, and appear on maps and figures as identifiers for the sampling points. Because the field records and computer data entry during sample receiving at the laboratory are done for the most part manually, errors in sample ID recording are common. *To reduce data management errors, sample numbers must be simple, short, and consecutive.*

An effective sample numbering system should be developed in the planning phase and documented in the SAP. Several practical rules for sample numbering are as follows:

Rule 1. Sample IDs that are too long and contain too much field information are cumbersome and impractical.

- An example of such sample ID is 'IR16-Area-25-MW25/01-W-12/02/01'. This sample number identifies a water sample collected at Installation Restoration Site 16, Area 25, Monitoring Well 25/01 on December 3, 2001. This sample ID may be too long to fit the appropriate field in the laboratory database or on a map and is needlessly complicated. The field information encoded in this ID is recorded otherwise in field notebooks, tracking logs, and the COC form. Sample numbers of this kind are unmanageable in the field and at the laboratory.

Rule 2. Sample IDs should not contain identifiers that are not unique.

- Project numbers, site IDs, and other information that is the same for all of the collected samples do not enhance sample IDs. For example, sample numbers 12345-IR16-S-001, 12345-IR16-S-002, and 12345-IR16-S-003 contain only one unique identifier, which is the sequential number at the end of each ID. The rest of the ID just fills the space.

Rule 3. Avoid sample IDs that look alike.

- Numbers, such as EX-01-0.75, EX-01-3.75, EX-02-0.75, EX-02-3.75, EX-03-1.0, EX-03-4.0, EX-04-0.75, EX-0.4-3.5, look alike and are susceptible to the transposition of digits in manual recording and in computer entry.

Rule 4. Number samples consecutively and as simply as possible.

- The best way to do this is to assign consecutive numbers to the samples as they are being collected. If different matrices are sampled, sample ID may include a matrix identifier. Examples of such sample numbers are S-001, S-002, W-003, W-004, WP-005, WP-006. The first two samples are soil, the next two are water, and the last two are wipes.

Rule 5. Groundwater samples collected from monitoring wells on a long-term monitoring program are usually identified by the well ID.

- Some databases will not accept samples with the same field ID. A unique identifier (a date or a sequential number) must be added to the monitoring well ID, for example, MW-1-120701, MW-2-120701, MW-3-120701 or MW-1-001, MW-2-002, MW-3-003.

Sample IDs are recorded on sample labels together with the information related to sampling and analysis, such as the following:

- Project name
- Sampling point ID
- Date and time of sampling

- Sampler's name
- Requested analysis
- Chemical preservation method

Appendix 10 shows an example of a sample label. Sample labels may be preprinted to avoid recording errors in the field, particularly for projects with a large number of samples. If filled out by hand in the field, sample labels must be legibly written with non-running ink and secured on the sample containers with clear tape.

Many environmental laboratories in the USA offer a turnkey automated sample tracking systems that include computerized COC forms and bar-coded sample labels.

3.2.3 Sample tracking

For any project, no matter how large or small, we should keep a *Sample Tracking Log*. A Sample Tracking Log ties together the following field and laboratory information that is vital for data review, data management, and report preparation:

- Field sample ID
- Laboratory sample ID
- Sample collection date
- Sample location (soil boring, depth, monitoring well, grid unit, etc.)
- Sampler's name
- Sample matrix
- Requested analysis
- Laboratory name (if several laboratories are used)
- Sample type (field sample, field duplicate, equipment blank)

Appendix 11 shows an example of a completed Sample Tracking Log. The information in the Sample Tracking Log is tailored to project needs, and the blank forms are prepared before sampling activities start. A sampling team member will enter the field information either by hand on a paper copy or electronically using a spreadsheet or a database. After laboratory data have been received, entering the laboratory sample IDs completes the Sampling Tracking Log. The COC form does not replace the Sample Tracking Log, as the latter combines the information on *all of the samples* collected for the project.

Practical Tips: Sample custody and tracking

1. The best way to track samples and manage the data is through a central database.

2. Make sure that you have a stock of blank COC forms prior to start of sampling.

3. Correct errors in the COC form by crossing out the wrong information with a single line and placing your initials next to it.

4. If a COC form becomes hard to read because of too many corrected errors, rewrite it completely.

Practical Tip: **Sample custody and tracking (continued)**

5. Each sample shipment should have a COC form that lists all the samples in the shipment.

6. Always indicate the turnaround time of analysis on the COC form.

7. Provide on the COC form the project contact name, address, and telephone and fax numbers.

8. If a COC is several pages long, number the pages in a format 'X of Y.'

9. When requesting project-specific MS/MSD analyses, collect extra sample volumes and designate the containers on the COC form as 'MS/MSD.'

10. Always identify the temperature blank on the COC form as a separate line item.

11. Always prepare a Sample Tracking Log; it facilitates efficient data review and report preparation.

3.3 SAMPLE PRESERVATION TECHNIQUES

A sample uniquely represents the sampled medium in a given location at the time of sampling. In a sense, it is a snapshot of the conditions within the medium, and these conditions are encoded in the analytical results. In order to ensure that analytical results truly represent the conditions at the time of sampling, all chemical and physical changes that may take place in a sample must be minimized. The following processes are known to alter the chemical state or the concentrations of contaminants in soil and water samples:

- Contamination from sample containers or preservation chemicals
- Degassing of dissolved oxygen, nitrogen, and methane from water samples
- Loss of carbon dioxide from water samples followed by a change in pH
- Adsorption of metals to container walls in water samples
- Absorption of atmospheric gases followed by oxidation and precipitation of metals in water samples
- Bacterial degradation of organic compounds in soil and water samples
- Volatilization of low boiling point organic compounds from soil and water samples
- Chemical and photochemical reactions in water samples

Preservation enables us to minimize the degradation of contaminant concentrations in samples and to assure the 'collected sample' representativeness. Although complete sample preservation is practically impossible, a significant reduction in contaminant degradation rates can be achieved by using proper sample containers; preserving samples with chemicals; keeping them refrigerated; and by observing the holding time requirements.

Based on the their knowledge of contaminant degradation in samples, the EPA and other professional organizations involved in sampling and analysis derived the requirements for container types and materials, sample preservation, and holding time. These requirements for soil and water samples are summarized in Appendices 12 and 13.

3.3.1 'Sample containers'

An important element in assuring the 'collected sample' representativeness is the use of sample containers that sustain sample integrity. We collect water samples for VOC analysis in 40-ml VOA vials with screw-on PTFE-lined silicon rubber septum caps. The PTFE liner of the septum should always face down to contact the water. Water samples for SVOC analyses are collected into 1-liter amber glass bottles with PTFE-lined lids. Samples for metal and inorganic parameter analysis are placed either into high-density polyethylene (HDPE) or glass bottles.

Soil samples for VOC analysis are collected into airtight coring devices or into preserved VOA vials. Soil samples for SVOC, metal, and inorganic parameter analyses are collected into brass or stainless steel core barrel liners, acrylic liners, or into glass jars with PTFE-lined lids. The liners are capped with PTFE sheets and plastic caps.

As a rule, we should use only precleaned sample containers certified by the manufacturer. Manufacturers clean glass and HDPE containers for water samples according to the EPA specifications and provide a certificate of analysis with each case of containers. Metal liners for core barrels and split spoons are usually not precleaned by the manufacturer, but they may be precleaned by the distributor. If precleaned liners are not available or the level of their cleanliness is questionable, we should decontaminate them prior to sampling. In this case, we may verify the effectiveness of the cleaning procedure by collecting a rinsate sample from a cleaned liner and analyzing it for the contaminants of concern. A decision on analyzing the liner rinsate blank should be solely based on the nature of the contaminants of concern and the project DQOs.

3.3.2 Sample preservation

We use a variety of preservation techniques in order to minimize the degradation of contaminant concentrations in soil and water samples. Preservation methods stem from the chemical nature of each contaminant and from the rate at which it will undergo irreversible changes in a sample. Preservation is achieved by the application of one or more of the following techniques:

- Storing samples at 2–6°C to reduce volatilization and bacterial degradation
- Adding acid to water samples to stop bacterial activity
- Adding acid or base to water samples to change the pH and thus preserve the chemical state of a contaminant
- Using bottles made of opaque (amber-colored) glass to reduce photochemical reactions in water samples
- Placing samples into containers made of materials compatible with the sampled medium and the contaminant
- Eliminating headspace in containers with samples for VOC analysis to minimize volatilization

Every analytical method has its own container and preservation requirements for each matrix. Containers for water samples that are precleaned and preserved with chemicals according to the EPA protocols are available from various manufacturers. We may also obtain them from the laboratory, which will provide analytical services for the project. The container cleanliness and the purity of preservation chemicals must be of the highest quality in order not to introduce any extraneous contamination into the samples.

If preserved containers for water samples are not available, we should obtain small quantities of preservation chemicals from the laboratory and preserve samples in the field immediately after they have been collected. Only laboratory-grade, high purity chemicals, which have been certified as free of contaminants of concern, should be used as preservatives.

Common to nearly all analyses is preservation with refrigeration at 2–6°C, a practice, which minimizes the volatilization of organic compounds with low boiling points and the bacterial degradation of most organic compounds. That is why we must place samples on ice immediately after they have been collected, ship them in insulated coolers with ice, and keep them refrigerated until the time of analysis. Water samples collected for metal analysis and preserved with nitric acid are an exception to this rule as they may be stored at room temperature. The addition of methanol or sodium bisulfate solution to soil collected for VOC analysis is the only chemical preservation techniques ever applied to soil samples.

Preservation methods for water samples are diverse and depend on the chemical nature of the contaminants. Several analytical methods require chemical preservation of water samples with acid, base or other chemicals as shown in Appendix 13. Typical chemical preservation techniques for water samples and the underlying chemical reasons are as follows:

- Samples for **VOC**, **TPH**, and **oil and grease** analyses are preserved with acid to destroy bacteria and stop the bacterial degradation of aliphatic and aromatic compounds.
- When exposed to oxygen in air, **dissolved metals** precipitate in unpreserved water as oxides or hydroxides. This process is prevented by the addition of 1:1 nitric acid solution to water samples. Metals in water with the pH of less than 2 will be permanently dissolved in a form of highly soluble nitrates.
- **Cyanide** species are reactive and unstable, and the addition of sodium hydroxide (in pellets or as a strong solution) stabilizes them as simple and complex alkali cyanides. Sodium hydroxide also prevents the release of the toxic hydrogen cyanide gas.
- Free chlorine destroys **cyanide** compounds. Water samples collected for cyanide analysis and suspected of containing chlorine are amended with ascorbic acid to remove chlorine.
- The addition of sodium thiosulfate to samples collected for **VOC** analysis removes residual chlorine that may be present in drinking water as an artifact from water disinfection process.
- **Ammonia**-containing samples are preserved with concentrated sulfuric acid to convert ammonia into the stable ammonium ion.

- Samples collected for ***total phenolics*** analysis by distillation are preserved with sulfuric acid to stop bacterial degradation and oxidation.
- ***Sulfide*** ion is preserved in the form of zinc sulfide, a precipitate that forms upon the addition of zinc acetate and sodium hydroxide to water samples.

Practical Tips: Sample preservation

1. Familiarize yourself with the sample container, preservation, and sample volume requirements for each sampling event.

2. Use only precleaned, preserved water containers that are appropriate for the contaminants of concern.

3. Remember that improper preservation may invalidate samples. Verify the container preservation before filling them with samples: the name of the preservation chemical must be clearly stated on the container label.

4. Preserve samples collected for metal analysis with 5 ml of 1:1 nitric acid solution.

5. Remember that cyanide must not come in contact with acid because the addition of acid to samples containing cyanides will produce deadly hydrogen cyanide gas.

3.3.3 Holding time

An important consideration in sample preservation is the observance of holding time, which is defined by ASTM (ASTM, 1987) as follows: '***The period of time during which a water sample can be stored after collection and preservation without significantly affecting the accuracy of analysis.***'

The EPA's similar definition of holding time (EPA, 1997a) sheds more light on the influence of holding time on data usability: '***Holding time—The period of time a sample may be stored to its required analysis. While exceeding the holding time does not necessarily negate the veracity of analytical results, it causes the qualifying or "flagging" of any data not meeting all of the specified acceptance criteria.***'

We will discuss the significance of holding time for data quality and the application of data qualification or 'flagging' as a data evaluation tool in Chapter 5. In this chapter we will discuss the importance of holding time as a preservation technique.

Observing the holding time for extraction and analysis is a universal measure to minimize the undergoing changes in soil and water samples. To control the risk of irreversible chemical and physical changes affecting contaminant concentrations, each class of contaminants has a defined technical holding time listed in Appendices 12 and 13, which depends on the rate of contaminant degradation in soil and water samples.

The CLP SOW makes a distinction between the ***technical*** holding time and the ***contract*** holding time, which is defined as the validated time from sample receipt (VTSR) at the laboratory. The VTSR is 2 to 4 days shorter than the technical holding time. This contract condition releases the laboratory of the responsibility for the missed technical holding time due to the sampling contractor's delay in shipping the samples to the laboratory. When the technical holding time is exceeded while the VTSR holding time is met, data quality may be affected.

The EPA-recommended holding time for several analyses in water samples may be excessively long. For example, the maximum holding time for chemical oxygen demand (COD) recommended in the Standard Methods is 6 hours and for biochemical oxygen demand (BOD) it is 7 days (APHA, 1998). The EPA-recommended holding time for COD and BOD is 48 hours and 28 days, respectively (EPA, 1983). The Standard Methods recommend a 7-day holding time for VOC analysis in water, whereas it is 14 days in EPA methods. These ambiguities raise questions as to the scientific basis that guided the holding time selection. The consensus among environmental professionals is that the holding times for high molecular weight analytes, such as PCBs, polynuclear aromatic hydrocarbons (PAHs), organochlorine pesticides, and dioxins, could be extended without adversely affecting the data quality. Organic and inorganic parameters susceptible to degradation in storage, such as VOCs, COD, BOD, total organic carbon, alkalinity, and some others, should be analyzed for as soon as possible after sampling, especially if low-level concentrations are a matter of concern.

3.4 SAMPLE PACKAGING AND SHIPMENT

After the samples have been collected, the sampling team should ship them to the laboratory in the most expedient manner. If samples cannot be shipped on the day of collection, they should be kept on ice or under refrigeration overnight.

The mode of sample delivery to the laboratory should be selected during the project planning phase based on the type of samples, their holding time, the distance to the laboratory, and the available transportation options. If a local laboratory with a courier service is within a reasonable driving distance from the project site, the sampling team should make arrangements for sample pickup at the site on the sampling days. If local laboratories are not available, samples should be shipped to a laboratory by an overnight delivery service.

Concerns for the VOC loss from water samples during transportation by air have been raised on occasion. Laboratory studies, however, have demonstrated that there is no discernable difference between VOC data obtained from analysis of water samples shipped by land and by air (Craven, 1998). The notion of unsuitability of air transportation for air samples collected in Tedlar® bags is also common. However, if Tedlar® bags are filled to 2/3 of their capacity, they survive air travel without leaking or bursting.

While in transport, the packaged samples must be kept on ice at 2–6°C; hence the need for insulated coolers. To prevent water damage from melting ice, the ice should be double-bagged in resealing bags. The greater the number of samples in a cooler, the more ice is needed to maintain the proper temperature range. Real ice (i.e. frozen water) used in copious quantities is the best choice. 'Blue ice' is not as effective and may be used only for very small shipments. Dry ice is not a widely used coolant because it freezes water samples and causes the containers to break.

A temperature blank placed with the samples inside the cooler is the best tool for determining the true temperature of the samples upon arrival to the laboratory. For samples that do not require refrigeration (Tedlar® bags with air, water samples for metal analysis), coolers are still the best shipping containers because they are light and durable.

There are two other important issues to consider when shipping environmental samples. The first issue is related to the fact that the United States Department of Transportation (DOT) and International Air Transportation Association (IATA) regulate the shipping of environmental samples by commercial courier services. These shipping authorities require that sample shipments be labelled according to type of samples and the nature of suspected contaminants. That is why we should familiarize ourselves with these labelling requirements, which are detailed in several DOT and IATA publications, and observe them at all times when shipping samples. Another issue concerns the shipping of soil samples collected overseas. Because the United States Department of Agriculture (USDA) does not allow uncontrolled import of foreign soil into the USA, analytical laboratories must obtain a special USDA permit for importing soil samples from abroad. We need to obtain a copy of this permit from the receiving laboratory ahead of time. When shipping foreign soil samples, we will attach this permit to the side of the shipping container for the examination by the customs officials.

To package samples for shipping, we will need the following supplies:

- Sturdy plastic or metal coolers without spouts or with spouts taped shut
- Resealing plastic bags of various sizes
- Ice
- Temperature blank
- Cushioning packing material, such as bubble wrap
- Clear strapping tape
- Warning labels ('This Side Up,' 'Fragile')
- Air bill, way bill forms or shipping labels
- Custody seals (optional)
- USDA permit (optional)
- Waterproof ink pen

To pack a cooler with samples, we follow these steps:

1. Prepare the labelled samples and complete the COC form.
2. Verify the label information against the COC form.
3. Place clear tape over the sample labels to protect them from moisture.
4. Place each sample container into individual resealing plastic bags and wrap them in bubble wrap. Several VOA vials containing the same sample may be placed in the same bag.
5. Place double-bagged ice on top of the bubble wrap on the bottom of the cooler.
6. Place samples and the temperature blank in the cooler, place bags with ice between the samples and on top of the samples.
7. Place additional bubble wrap on top of the samples to prevent them from moving and shifting if the cooler is overturned.
8. Record the air bill or waybill number on the COC form, sign the form, place it in a resealing plastic bag, and tape it onto the inside surface of the cooler lid.
9. Close the lid and tape it with strapping tape. Place the package orientation labels and other warning labels as required by the DOT or IATA on the sides of the cooler.

10. When shipping foreign soil samples, tape a copy the USDA permit for importing soil in a clear plastic to the side of the cooler for the examination of the customs official.
11. If required by the project SAP, place dated and signed custody seals on the front and on the back side of the lid to seal cooler.
12. Complete air bill or waybill or shipping label; tape it to the top of the cooler. Transfer the cooler to the shipping company.

Practical Tips: Sample packaging and shipping

1. To minimize the number of trip blanks and the cost of their analysis, segregate water samples for VOC analysis in one cooler.

2. For long-term projects, place an icemaker and sample refrigerators on the project site.

3. Prior to start of sampling, find out where the nearest overnight delivery service office is located. If possible, arrange sample pickup from the project site.

4. Allow yourself plenty of time for sample packing: it usually takes longer than expected to properly pack a cooler.

5. Use plenty of cushioning material to prevent glass bottles from breaking in transport.

6. Remember that a cooler with samples usually does not keep the inside temperature of 2–6°C for more than 24–48 hours; always ship samples for next day delivery.

7. Familiarize yourself with shipping regulations and shipping container labelling requirements as you may be personally liable for improper shipment of hazardous materials (i.e. acid and base-preserved samples or PCBs).

8. When shipping samples to the USA from abroad, request from the laboratory the USDA permit for importing soils. Display the permit by taping it in a clear plastic bag to the side of the cooler.

9. Request morning delivery from the courier service to give the laboratory plenty of time to receive the samples.

10. When sampling on Friday, coordinate sample receipt by the laboratory on Saturday.

11. Be aware of the holding time requirements; do not store samples on the project site without a cause. Make a habit of shipping samples to the laboratory on the day of collection.

3.5   SOIL SAMPLING DESIGNS

This chapter provides guidance on several basic aspects of soil sampling, such as systematic sampling on a grid system, sample compositing, and sampling for specific chemical parameters, such as PCBs and VOCs.

3.5.1   Laying out a sampling grid

In the course of systematic soil sampling, we use grid patterns for identifying the sampling points. The size of the sampling grid and the locations of sampling points within each grid unit are defined in the SAP. A well-prepared SAP will contain a figure showing the map of the site with the sampling grid, the sampling points, and the reference landmarks clearly identified.

We use sampling grids mainly for surface soil sampling, however, it is not unusual to place soil borings and collect subsurface samples on a grid pattern. Grids are also used for soil sampling from the bottom and sidewalls of excavation pits and trenches.

The sampling grid design is based on the project DQOs and is described in the SAP. The first sampling point is randomly selected during the optimization of the sampling design in the DQO process Step 7. However, quite often the SAP does not provide enough detail on the sampling grid orientation and exact positioning, and the only information available to the field crew is the size of the grid unit and the total number of samples to be collected. The field crew then will select the coordinates of the first sampling point within the grid by using a random number generator and will start building the grid rows from this point. In this case, the total number of samples may be different from the proposed number of samples. For sampling surface areas or shallow excavations of irregular shape, the sampling grid may start outside the area to be sampled.

Sampling grids can be square, rectangular, or hexagonal (triangular) with the grab samples collected from each grid nodule. Depending on the project DQOs, grab samples may be composited for analysis.

The most efficient way to lay out an accurate sampling grid is to employ a land surveyor who will identify the sampling points and record the northing and easting coordinates with the assistance of a Global Positioning System (GPS). Differential GPS has an accuracy of 1–3 meters that may be sufficient for grids with large unit size. If this option is not available, we may construct the grid using simple tools available from a hardware store.

We need the following field equipment for laying out a sampling grid:

- Tape measure
- Compass
- Stakes or surveyor flags
- Surveyor spray paint
- String or rope
- Field notebook
- Pen
- Calculator
- Three 3–4 feet (ft) long poles (for hexagonal or triangular grids only)
- Photo camera

3.5.1.1 Square grid
Figure 3.3 shows an example of a square grid with a grid unit length of 20 ft. To build a square grid, we will follow these steps:

1. Identify a permanent landmark at the site (a building, a utility pole).
2. Determine your bearings with a compass.
3. From the SAP, determine the size of the grid and its orientation relative to the permanent landmark.

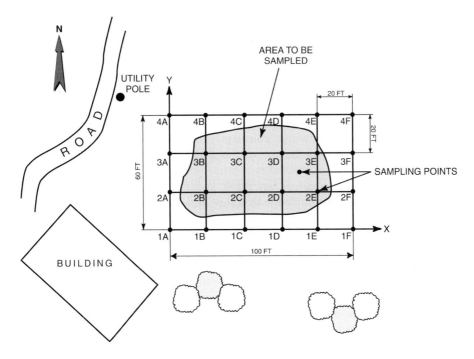

Figure 3.3 Square sampling grid.

4. Referencing to the permanent landmark and using a compass for orientation, delineate the grid perimeter with stakes and string. (In Figure 3.3, the area is 100 ft long and 60 ft wide).
5. Divide the perimeter of the area into grid unit lengths per the SAP and stake these points. (In Figure 3.3, the points are identified as 1A to 1F on the southern boundary; 1A to 4A on the western boundary; 4A to 4F on the northern boundary; and 1F to 4F on the eastern boundary.)
6. Mark the string along the southern boundary with paint at the sampling points.
7. Move the marked string by one grid unit length to points 2A and 2F and stretch it parallel to the first row of sampling points. Stake the second row of sampling points. (Points 2A through 2F in Figure 3.3.)
8. Move the marked string by another grid unit length and stake the third row of samples (Points 3A through 3F in Figure 3.3.)
9. Make a drawing of the grid in the field notebook. The completed grid extends beyond the boundaries of the irregular area to be sampled as shown in Figure 3.
10. Collect samples and document sampling according to procedures described in the SAP; take a photograph if unusual conditions are encountered.
11. Survey the sampling points, if required by the SAP.

There are a number of options for sampling on a grid system. For example, when justified by the project DQOs, instead of collecting samples in the grid nodules, we may collect one random sample within the boundaries of each grid unit or collect several samples from each grid unit and composite them for analysis.

3.5.1.2 Hexagonal grid

Hexagonal (triangular) grid construction is best achieved using a simple but an amazingly effective device, which is a flexible equilateral triangle made of three poles and a rope. To build a hexagonal grid, similar to the one shown in Figure 3.4, we follow these steps:

1. Identify a permanent landmark at the site (a building, a utility pole).
2. Determine your bearings with a compass.
3. From the SAP, determine the size of the grid and its orientation relative to the permanent landmark.
4. Referencing to the permanent landmark and using a compass for orientation, delineate the area to be sampled with stakes or flags and string. (In Figure 3.4, the area is 100 ft long and 60 ft wide.)
5. Using three poles and a string, make an equilateral triangle with the length of the string between the poles (the side of the triangle) of 20 ft as shown in Figure 3.4. We will use this triangle as a measuring device.
6. Using a tape measure, divide the northern and southern boundaries of the grid into 20-ft increments. Stake sampling points 1 to 6 and 23 to 28 as shown in Figure 3.4.

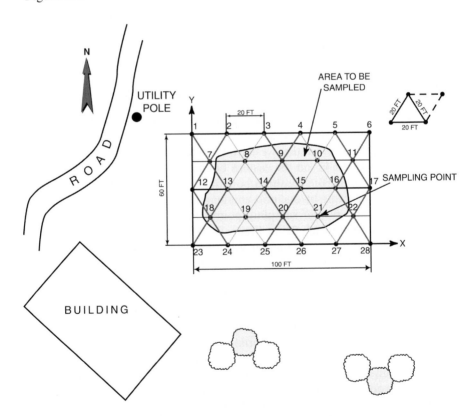

Figure 3.4 Hexagonal sampling grid.

7. Place two poles of the pole triangle on the border of the delineated area (points 1 and 2 in Figure 3.4); place the third pole inside the delineated area (point 7 in Figure 3.4); stretch the string tightly. Stake point 7.
8. Without moving the pole in points 7 and 2, move the pole from point 1 to point 8. Stake point 8.
9. Without moving the poles in points 7 and 8, move the pole from point 2 to point 13. Stake point 13.
10. Without moving the poles in points 8 and 13, move the pole from point 7 to point 14. Stake point 14.
11. Without moving the poles in points 8 and 14, move the pole from point 13 to point 9. Stake point 9.
12. Continue these steps until entire grid is established.
13. Make a drawing of the grid in the field notebook. The completed grid extends beyond the boundaries of the area to be sampled as shown in Figure 3.4.
14. Collect samples and document sampling according to the procedures described in the SAP; take a photograph if unusual conditions are encountered.
15. Survey the sampling points, if required by the SAP.

Once this grid has been laid out, we may collect grab samples from grid nodules or choose composite sampling on a pattern as shown in Figure 3.5, where each square represents a six-point composite.

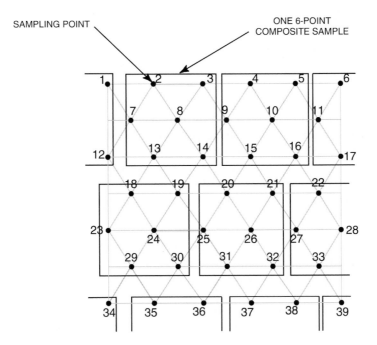

Figure 3.5 Composite sampling on a hexagonal grid.

3.5.2 PCB sampling designs

Requirements for the collection of soil data for cleanup verification of PCB-contaminated sites are regulated by the TSCA Subparts N and O (EPA, 1998c). These detailed and exacting requirements are summarized in this chapter in the form of procedures.

As a general TSCA rule, at each separate cleanup site we must take a minimum of *three* samples for each type of bulk PCB remediation waste or porous surface. There is no upper limit on the number of samples. Bulk PCB remediation waste includes, but is not limited to the following non-liquid PCB-contaminated media: soil, sediments, dredged materials, mud, sewage sludge, and industrial sludge. Example 3.1 illustrates the application of this rule.

Example 3.1: Samples of bulk remediation waste for PCB analysis

A project consists of the removal of PCB-contaminated media at three different cleanup sites: a loading dock, a transformer storage lot, and a disposal pit. There are several different types of bulk PCB remediation wastes at the three sites that must be identified and sampled separately as follows:

- Loading dock—concrete (3 samples) and clayey soil (3 samples).
- Transformer storage lot—oily soil (3 samples), clayey soil (3 samples), and gravel (3 samples).
- Disposal pit—sandy soil, clayey soil, oily soil, industrial sludge, and gravel; we will collect a minimum of 3 samples from each of them.

A total number of samples collected for the project will, therefore, be at a minimum 30.

The TSCA requires that all PCB results be reported on a dry weight basis and that the records be kept on file for 3 years. It is interesting to note that TSCA employs the metric system, which is rarely used in the USA.

3.5.2.1 Cleanup verification on a grid for multiple sources

For site cleanup verification after the removal of PCB contamination, the TSCA provides specific requirements on the size and the orientation of the grid and the depth at which samples are to be collected. Figure 3.6 shows the grid used for multiple source PCB cleanup verification according to these requirements.

The procedure of the grid layout is the same as described for a square grid in Chapter 3.5.1.1 with the following modifications:

1. Determine the center of the area and orient the grid axes on north–south and east–west directions using the center of the area as the origin.
2. Overlay a square grid system with a grid unit length of 1.5 meter (m) over the entire area to be sampled.
3. If the site has been recleaned after initial cleanup verification and samples are being collected the second time, reorient the grid axes by moving the origin 1 m to the north and 1 m to the east.
4. Collect grab samples from the grid nodules at the depth of 7.5 centimeters (cm).

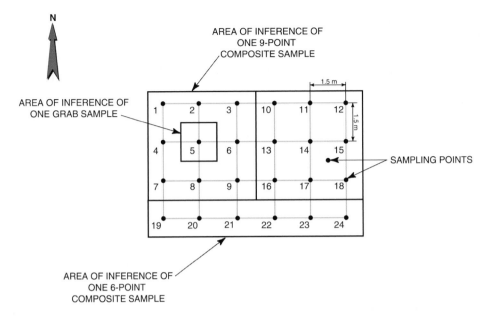

Figure 3.6 Grid design for PCB sampling—multiple sources.

Analytical results for an individual sample point apply to the sampling point itself and *an area of inference*, which forms a square around the sampling point with the side length equal to the grid unit length as shown in Figure 3.6. The area of inference from a composite sample is the total of the inference areas of the individual samples included in the composite.

This sampling design is applicable to the sites with multiple point sources (multiple spills) or for sites where the source of contamination is unknown. The TSCA recommends sample compositing for such sites. Samples, collected from the nodules of adjoining grid units (1 through 9, 10 through 18, and 19 through 24 in Figure 3.6), may be composited. A maximum number of grab samples allowed to be composited is nine. Additional grab samples may be needed to fully characterize the areas of the site where compositing is not feasible.

Either field or laboratory personnel prepare composite samples by mixing equal aliquots of grab samples and taking an aliquot of a well-mixed composite of sufficient weight for laboratory analysis. Grab samples selected for compositing must be of the same type of bulk PCB remediation waste; the TSCA does not allow compositing of different types of PCB waste.

For a small site with multiple sources, we may use the same compositing design for samples collected on a smaller grid with a smaller grid unit length. For smaller grids, the area of inference for each composite will be also smaller. The same design may be applied to sampling for the purpose of site characterization with one exception: a 3-m grid is used instead of a 1.5-m one.

3.5.2.2   Random surface sampling

For small or irregularly shaped cleanup sites where a square 1.5 m grid will not result in a minimum of three sampling points, the TSCA offers two options. The first option is to use a smaller grid as appropriate for the size of the site and the procedures described in Chapter 3.5.2.1. The second option offers the following coordinate-based random sampling scheme:

1. Beginning in the southwest corner of the area to be sampled, measure the maximum north–south and east–west dimensions of the total area and delineate its boundaries as a square or a rectangle.
2. Designate the west and the south boundaries of the area as the reference axes of a square grid system. Using a random number generator program, determine a pair of random coordinates that will locate the first sampling point within the area to be sampled. The first coordinate of the pair is a measurement on the north–south axis; the second coordinate is a measurement of the east–west axis. Collect the samples at the intersection of these north–south and east–west lines drawn through the measured points on the axes. If this intersection falls outside the area to be sampled, select another coordinate pair.
3. Continue to select coordinate pairs until a minimum of three random sampling points are located for each type of bulk PCB remediation waste.
4. Collect samples to a maximum depth of 7.5 cm.
5. Analyze all samples as grab or composite samples. For a small site, the result of a three-point composite analysis may represent the entire site.

3.5.2.3.   Single point source

The TSCA specifies a composite sampling design applicable to a single point source of contamination, such as a discharge into a large containment area (a pit or evaporation pond) or a leak into the soil from a transformer, a single drum or tank. Compositing in this case comes in two stages: *initial* compositing and *subsequent* compositing from concentric square zones around the initial compositing area as shown in Figure 3.7.

The steps for laying out a composite sampling grid for a single point source of PCB contamination are as follows:

1. Determine the center of the area to be sampled, and using it as the grid origin, lay out the first four squares 2.25 m by 2.25 m large as shown in Figure 3.7. Follow the procedures on the grid orientation described in Chapter 3.5.2.1. The first four squares form the initial compositing area, which is shaded in Figure 3.7. The overall sampling grid will be built on the same axes as these squares.
2. Collect 9 grab samples from the center of the area and the perimeter of the 4 grid units in the initial compositing area. These samples, numbered 1 through 9 in Figure 3.7, will be composited for analysis.
3. On each axis step out 1.5 m from the boundaries of the initial compositing area. Using these marks, delineate the first subsequent compositing area around the initial compositing area by building a square with the sides that are 7.5 m long as shown in Figure 3.7.

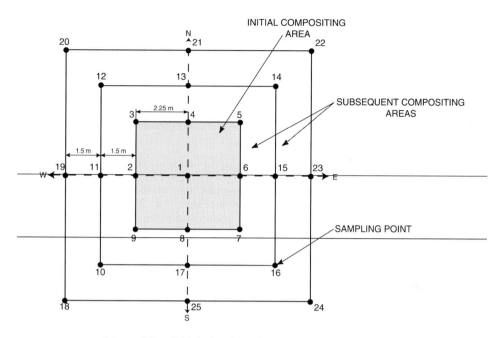

Figure 3.7 Grid design for PCB sampling—single source.

4. Collect 8 grab samples from the perimeter of the first subsequent compositing area (samples 10 through 17 in Figure 3.7).
5. To delineate the second subsequent compositing area, step out another 1.5 m on each axis and delineate a square with the sides that are 10.5 m long. Collect another 8 grab samples from the perimeter of the second subsequent compositing area (samples 18 to 25 in Figure 3.7).
6. Continue laying out the subsequent compositing areas until the total site area is covered. Collect 8 grab samples from the perimeters of each subsequent compositing areas.

As with the other PCB sampling designs, the samples are collected at the maximum depth of 7.5 cm and only similar types of waste may be composited for analysis. Additional grab samples may be also required to fully characterize the area.

Practical Tip: Laying out a grid

1. Building a sampling grid without the help from a land surveyor requires an effort of two to three persons.

2. To be efficient while sampling, locate and stake the sampling points prior to sampling.

3. Make a copy of the sampling grid map from the SAP and use it in the field for recording of actual measurements and sampling points.

---

continues

> **Practical Tip: Laying out a grid (continued)**
>
> 4. If the SAP does not contain the grid maps, make accurate drawings of the sampled areas and document the exact positions of the sampling points.
>
> 5. Select landmarks that are easily identifiable, unique, and truly permanent.
>
> 6. Do not spray-paint the sampling points as the paint may contain solvents that interfere with analysis.
>
> 7. Label stakes or surveyor flags with sampling point or sample IDs.
>
> 8. Always number the grid system units. If samples are collected in the *grid nodules*, number each nodule numerically as shown in Figure 3.4. If random samples are collected within *each grid unit*, use an alphanumeric numbering system to identify each grid unit as shown in Figure 3.3.
>
> 9. Photograph the site to show the sampling grids and permanent landmarks.
>
> 10. Compositing in the field is often impractical as it requires additional resources. The laboratory will composite samples for analysis if given proper instructions.

3.5.3 Sampling from excavation pits

In the course of remediation projects we often need to sample from the bottom and sidewalls of excavation pits. A variation of excavation pit sampling is sampling from test pits and trenches. Test pits are small excavations usually made with a backhoe for a purpose of collecting a subsurface soil sample. A trench is a long and narrow excavation originating from a pipeline removal or placed with earth moving equipment for exploratory purposes.

There are two methods for sampling soil from excavation pits and trenches: systematic and judgmental. Systematic sampling is appropriate for large excavated areas with random spatial contaminant distribution. A sampling grid is overlaid on the bottom and sidewalls of a pit or a trench and samples are collected at regular intervals. Samples from small excavation pits and test pits are usually collected based on judgment.

A separate group of excavation pits are petroleum UST pits with their well-defined source of contamination. In the USA, many states regulate petroleum fuel UST removal and site remediation and establish action levels for petroleum fuels and their constituents in soil and groundwater. Many states also regulate the associated sampling of soil and groundwater. For example, the guidelines of the *Technical Guidance Document for Underground Storage Tank Closure and Release Response* (SHDH, 1992), regulate all UST removal activities in the State of Hawai'i. According to these guidelines, the number and the locations of samples depend on the size of the tank as shown in Table 3.1. UST sampling and analysis guidelines exist in the states of California, Arizona, Alaska, and some others.

The regulated UST pit and pipeline sampling is judgmental sampling. The selection of sampling points is biased towards the most probable sources of leaks, such as the

Table 3.1   Underground storage tank pit sampling guidelines[1]

Tank size	Number of soil samples	Number of water samples	Location of soil samples
Groundwater not present			
Less than 1,000 gallons	1	None	Fill or pump end of tank
1,000–10,000 gallons	2	None	At each end of the tank
Over 10,000 gallons	3 or more	None	Ends and middle of the tank along the longitudinal axis
Pipeline	1 or more	None	Every 20 linear feet, preferably at piping joints, elbows, and beneath dispensers
Groundwater in the pit			
Less than or equal to 10,000 gallons	2	1	From the wall at the tank ends at the soil/ground water interface
Over 10,000 gallons	4	1	From the wall at tank ends at the soil/groundwater interface
Pipeline	1 or more	None	Every 20 linear feet, preferably at piping joints, elbows, and beneath dispensers

[1] SHDH, 1992.

fill or pump end of the tank or the joints of the pipeline. This approach may be used for sampling of the excavation pit that housed a UST with any liquid material, not necessarily a petroleum product. We may also use a combination of systematic and judgmental sampling for USTs with large capacities (exceeding 10,000 gallons). Figure 3.8 shows an example of a rectangular large capacity UST excavation pit with a sampling grid overlaid on the floor and the sidewalls.

Sampling from an excavation pit, a trench or a test pit is a combined effort of the earth moving equipment operators, field labor crews, and sampling teams. Before a pit is ready for sampling, field crews may need to perform shoring and grading to make the entry safe. Pits may require dewatering, i.e. removing accumulated groundwater, rainwater, or perched water, by pumping the water from the pit into a holding vessel. Light nonaqueous-phase liquids (LNAPLs), such as petroleum fuel products, may be observed on the surface of the water in the UST excavation pits, either as sheen or as a layer of free product. If the latter is present, the field crews may remove it as a separate phase prior to dewatering the pit.

We must remember that entering a deep excavation might not be safe. According to the United States Occupational Health and Safety Administration (OSHA), confined or enclosed space means 'any space having a limited means of egress, which is subject to the accumulation of toxic or flammable contaminants or has an oxygen deficient atmosphere. Confined or enclosed spaces include, but are not limited to, storage

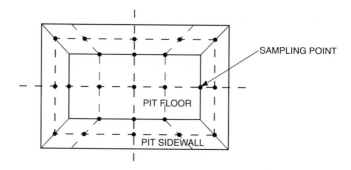

Figure 3.8 Sampling from an underground storage tank pit.

tanks, process vessels, bins, boilers, ventilation or exhaust ducts, sewers, underground utility vaults, tunnels, pipelines, and open top spaces more than 4 feet in depth such as pits, tubs, vaults, and vessels,' (OSHA, 1997). The dangers of confined space may include the presence of toxic, explosive or asphyxiating atmospheres or engulfment from small particles, such as dust. Confined space entry may require a permit; and every confined space, including excavations more than 4 ft deep, must be made safe for the workers. Only after the pit is made safe for entering, the sampling team may enter the pit and collect soil samples from the bottom and sidewalls. If a pit cannot be entered, an excavator is used to retrieve the soil from the sampling points, and samples are collected from the excavator bucket brought to the ground surface level outside the pit.

Overlaying a grid system over a deep excavation pit may not be always possible. In this case, we need to delineate a regularly shaped area around the pit and to mentally project a grid system over the pit as illustrated in Figure 3.9.

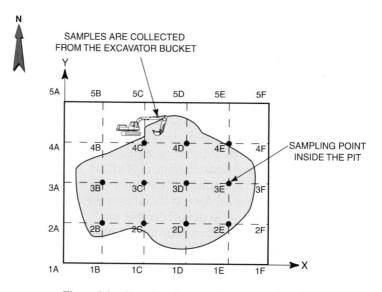

Figure 3.9 Sampling from a deep excavation pit.

To overlay a sampling grid over an excavation pit, we follow these steps:

1. Measure the pit dimensions (length, depth, and width) and record them in a field notebook. Several measurements will be needed if a pit is of irregular shape or of varying depth.
2. If water is present in the pit, sample the water for analysis with a bailer or a long-handled ladle. If an LNAPL (free product) is present, do not sample the water. Document the presence of the LNAPL and sample it, if required.
3. Photograph the pit to document the presence of water and the LNAPL.
4. Dewater the pit, if required.
5. If a pit is safe for entry, lay out a sampling grid inside the pit using procedures described in Chapters 3.5.1 and 3.5.2.
6. Collect undisturbed soil samples at the bottom and sidewalls of the pit.
7. If a pit is not safe for entry, delineate on the ground level the outer boundaries of the sampling grid with stakes and string. The dimensions of the outer boundaries should always be larger than the pit dimensions as shown in Figure 3.9.
8. Divide the boundaries into grid unit lengths and stake them or mark them on the ground with surveyor spray paint. These will be the X and Y coordinate markers for each sampling point.
9. Mentally project the sampling grid to extend it from the grid boundaries onto the pit. Direct the excavator operator to retrieve a bucket of soil from each imaginary sampling point.
10. Once the operator brings the full bucket to the ground surface, collect a sample from an undisturbed portion of soil in the bucket.

3.5.4 Stockpile sampling

Sampling of stockpiled soil is frequently conducted at remediation projects. The accessibility of soil for sampling usually depends on the size of the stockpile: the larger the stockpile, the more difficult the sampling. That is why the stockpile size is a key factor in the design of a stockpile sampling strategy.

Small and medium size stockpiles are usually sampled using a three-dimensional simple random sampling strategy. Grab samples are collected from the sampling points at different locations and depths within the stockpile. If the sampling is limited to certain portions of the stockpile, then the collected sample will be representative only of those portions, unless the soil is homogenous. Grab samples may be analyzed individually or as composites.

A large stockpile may be characterized using a systematic sampling approach. The stockpile is divided into segments of equal volume and each segment is sampled at random. The size of the segment is often mandated by regulatory guidance or by the disposal facility requirements for stockpile sampling. To reduce analytical costs, grab samples representing each segment of a given volume may be composited for analysis.

Whether the intended use of soil data is the determination of its suitability as backfill or the disposal characterization, we should always determine how many samples are required to characterize a stockpile with a certain level of confidence. The minimum number of samples will depend on the size of the stockpile and the applicable action level. If soil is to be used as clean backfill, a statistical evaluation of the required number of samples, similar to one shown in Example 2.3, is necessary. In practice, budget limitations usually overrun the sampling considerations, and as the result, stockpile characterization is often not conducted properly.

The *Stockpile Statistics Worksheet*, shown in Appendix 14, is a useful supplement for determining a minimum number of samples for proper stockpile characterization, if financial restrictions are not an issue.

To use the worksheet, we must first collect a number of stockpile characterization samples based either on regulatory guidance or on a specific sampling scheme with a certain budget in mind. Analytical results for these samples are entered into the worksheet as the mean concentration; the 95 percent upper confidence interval is calculated. A minimum number of samples is calculated in Worksheet Step 11; Step 12 allows determining if a sufficient number of samples have been collected. If not, additional sampling may be necessary. Examples 2.3 and 5.13 illustrate these calculations.

3.5.4.1   Determining stockpile volume

Smaller soil stockpiles are usually conical or pyramidal in shape. Large stockpiles are elongated and have an elliptical base and a semicircle, triangle, or a trapezoid face (cross-section). Figure 3.10 schematically presents a conical stockpile and an elongated stockpile with a trapezoid face.

To apply a grid system to a stockpile, we must first determine its volume using the following calculations.

Conical or Pyramidal Stockpile

1. Measure the height and the base parameters of the stockpile. The base parameters may be the length and width of a rectangular base or the radius of a circular base.
2. Calculate the base area (A).

Square or rectangular base:	$A = L \times W$	—L is length of one side of the stockpile base
		—W is length of the other side
Circular base:	$A = \pi \times r^2$	—r is radius of the circle
		—$\pi = 3.14$
Elliptical base:	$A = 1/2 \times \pi \times a \times b$	—a is length of the minor axis
		—b is length of the major axis
		—$\pi = 3.14$
Triangular base:	$A = 1/2 \times B \times h$	—B is base length of the triangle
		—h is height of the triangle.

3. Calculate volume of the stockpile (V). For all base shapes use the following equation:

Volume:	$V = 1/3 \times A \times H$,	—A is the base area
		—H is the height to the apex of the stockpile.

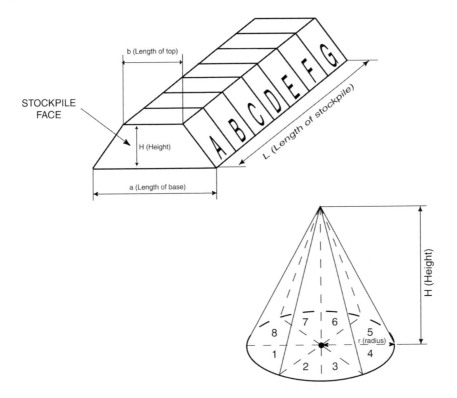

Figure 3.10   Stockpile sampling grid systems.

4. If the base is irregularly-shaped, divide it into regularly shaped areas, calculate the volume of each subsection with the regularly-shaped bases, and add the volumes together.

Elongated Stockpile

1. Determine the shape of the stockpile face (semicircle, trapezoid, or triangle). Measure the stockpile length, and the width and height of its face.
2. Calculate the area of the face (C).

Semicircular face:	$C = 1/2 \times \pi \times r^2$	—r is radius of the circle
		—$\pi = 3.14$
Trapezoid face:	$C = 1/2 \times (a + b) \times H$	—a is length of the base
		—b is length of the top
		—H is height of the trapezoid
Triangular face:	$C = 1/2 \times a \times H$	—a is base length of the triangle
		—H is height of the triangle

3. Calculate volume of the stockpile (V). For all shapes use the following equation:

Volume: $$V = C \times L$$ —C is the face area

                                                  —L is the length of the stockpile

4. If the stockpile face cannot be approximated with any of these geometrical figures, divide it into several regular shapes, and calculate the total volume as a sum of all volumes.

3.5.4.2   Stockpile sampling on a grid

Prior to sampling, the stockpile is divided into three-dimensional segments from which samples will be collected. An elongated stockpile is sliced like a loaf of bread (segments A through G), and a conical one is cut into wedges like a pie (segments 1 through 8) as shown in Figure 3.10. The size of the segment is often mandated by regulatory guidance for stockpile sampling. To sample a stockpile, we follow these steps:

1. Determine the volume of the stockpile as described in Chapter 3.5.4.1.
2. Determine the sampling frequency, i.e. how many cubic yards will be represented by one sample and how many grab samples will be collected.
3. Calculate the dimensions of each segment to be sampled and mark the segments along the stockpile base with stakes or flags.
4. Determine the sampling point locations within each segment with a random number generator. Sample each segment either manually or with an excavator or backhoe bucket. If the stockpile is stable and safe to climb on, use a hand auger or a hand trowel to obtain samples from below the surface at the top and sides of the stockpile.

Practical Tips: Stockpile sampling

1. One cubic yard equals 27 cubic feet.

2. One cubic meter equals 1.308 cubic yards.

3. One ton of soil has the volume of 0.75 cubic yards.

4. Soil density ranges from 80 to 130 pounds per cubic feet.

5. Before designing a sampling grid for an irregular stockpile, obtain as accurate measurements as possible, make an accurate drawing, and calculate the volume by breaking the stockpile into regular, geometrical subsections.

6. Stockpiles may be unstable and unsafe for climbing. Make sure that provisions for an excavator or a backhoe are made in such cases.

3.5.5 Sampling for volatile organic compounds

Sampling soil for VOC analysis is a separate field of knowledge. Many different sampling techniques exist today, which, if properly applied, will provide a relatively accurate representation of the true VOC concentration in soil. The limitations of each technique do not permit to select any one of them as a universal practice. For this reason, a combination of several techniques may be used to provide the best possible data.

Volatile organic compounds, which have the boiling points below 200°C, include the following classes of compounds:

- Aromatic and halogenated hydrocarbons and solvents
- Volatile petroleum fuels and solvents (naphtha, gasolines, jet fuels, mineral spirits)
- Ketones
- Acetates
- Nitriles
- Acrylates
- Ethers
- Sulfides
- Organic lead compounds (tetraethyl and tetramethyl lead)

The loss of VOCs from soil during sampling, shipping, storage, and even during analysis itself is a well-documented problem. For example, when exposed to air at room temperature, a soil sample will experience a 100 percent reduction in the trichloroethylene concentration within 1 hour (Hewitt 1996).

VOCs in soil exist in gaseous, liquid, and sorbed phases. The loss of VOCs from soil samples occurs mainly through volatilization and bacterial degradation. The gaseous phase (vapor) easily diffuses through soil pores and dissipates into the atmosphere instantly from the exposed surface. To compensate for this loss and to restore the equilibrium, liquid and sorbed VOCs volatilize and desorb, then diffuse through the pore space, and dissipate. Volatilization is particularly rapid when soil is aerated at ambient temperature, a process that takes place during sampling of disturbed soil.

Aerobic bacterial degradation of chemical contaminants takes place in soil with a proper balance of beneficial conditions (oxygen and moisture contents, ambient temperature, and nutrient supply) enabling the proliferation of heterotrophic bacteria. Certain bacteria metabolize and destroy chemicals as part of their nutritional intake, hence the term *bacterial degradation* or *biodegradation*. Aerobic bacteria attack the molecules that are easy to digest; straight chain aliphatic hydrocarbons are their first choice, followed by branched aliphatic and aromatic hydrocarbons. The process of biodegradation successfully removes many chemicals from environmental matrices. If favorable conditions exist, this process takes place *in-situ* in soil or groundwater as part of *natural attenuation*.

If one of the beneficial conditions for biodegradation is not present, the process will slow down or stop. Conversely, if the beneficial conditions are enhanced, biodegradation will proceed at a faster rate. Aeration of soil during sampling is one of such enhancements, which may lead to increased bacterial activity. Unlike rapid volatilization loss in disturbed soil, biodegradation loss is relatively slow. In 5 days of storage at room temperature, bacterial activity destroys about 85 percent of aromatic hydrocarbons, and in 9 days it removes them completely; under the same conditions, chlorinated hydrocarbons experience only a minor concentration loss (Hewitt, 1999a). Cold storage at 2–6°C slows down the biodegradation process. However, after a sample has been in cold storage for 14 days, benzene concentration in it may be reduced by as high 85 percent, and over 40 percent of toluene may be lost in 20 days (Hewitt, 1999a).

These facts demonstrate the significance of volatilization and biodegradation effects on the fate of VOCs in soil samples. These processes will inevitably take place unless special measures are implemented to stop their progress. A combination of preventive measures designed to minimize the VOC loss during sampling, storage, and analysis are implemented in the field and at the laboratory as part of EPA Method 5035, *Closed-System Purge-and-Trap and Extraction for Volatile Organics in Soil and Waste Samples.*

This method describes the procedures for closed-system soil sampling into special airtight coring devices or autosampler vials that are ***never opened after the sample has been collected***. Samples collected in this manner are analyzed with the purge and trap sample introduction technique and EPA Methods 8021 or 8260 (or their equivalents from other analytical series). An example of an airtight coring device is shown in Figure 3.11. It is the patented En Core™ sampler made by En Novative Technologies, Inc. This coring device is a sampling cartridge, which works like an airtight syringe. It has a barrel and a plunger; hermetic closures are achieved with o-rings made of inert material. The device is attached to a special sampling handle and pushed into soft soil. The plunger of the syringe moves up, allowing the soil to fill the barrel. Once the barrel is full, it is manually sealed with a hermetic cap. A sealed airtight coring device serves as a shipping container and may be used for sample storage at the laboratory. These devices may house 5 grams (g) or 25 g of soil.

For purge and trap analysis, soil is mixed with water to create slurry. The laboratory uses an extruding handle to transfer the samples from an airtight coring device into an autosampler vial with water prior to analysis. Samples may be also collected directly into autosampler vials with water. *After the samples have been collected, the vials are never opened*. Analytical laboratories use specially-designed purge-and-trap autosamplers to strip VOCs from the soil/water slurry with a combination of an inert gas purge, mechanical stirring, and the application of gentle heat (40°C).

EPA Method 5035 also addresses soil sample preservation in the field with methanol and sodium bisulfate and soil samples holding time requirements. A combination of the airtight device sampling technique, rigorous preservation, and holding time restrictions has been proven to produce accurate data on VOC concentrations in soil samples (Hewitt, 1995a; Hewitt, 1995b; Turiff and Reitmeyer, 1998; Hewitt, 1999a; Hewitt, 1999b).

EPA Method 5035 offers a variety of sampling options for soils of different lithological compositions and with different ranges of contaminant concentrations. Table 3.2 presents a summary of these options together with a variety of preservation

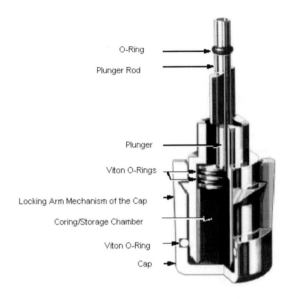

O-Ring
Plunger Rod

Plunger
Viton O-Rings
Locking Arm Mechanism of the Cap
Coring/Storage Chamber
Viton O-Ring
Cap

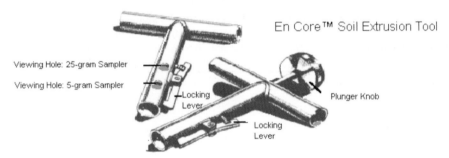

En Core™ T-Handle for Sampling

En Core™ Soil Extrusion Tool

Viewing Hole: 25-gram Sampler
Viewing Hole: 5-gram Sampler
Locking Lever
Plunger Knob
Locking Lever

Figure 3.11 En Core® sampler with attachable handles for sample collection and extrusion (from Hewlett, 1999a).

methods and holding time requirements. The sampling options presented in Table 3.2 are appropriate for soil with low VOC concentrations as they enable the laboratories to achieve the lowest PQLs for VOCs in soil.

When selecting a sampling option for VOC analysis in soil, we should consider the following factors:

- Suspected range of VOC concentrations at the site
- Lithological composition and the grain size of soil to be sampled
- Logistics of daily sample shipment from the project site to the laboratory that may affect the sample holding time
- Project budget

Table 3.2   Summary of soil sampling options for VOC analysis

Sampling method	Preservation	Holding time
Low detection limits—VOC concentrations below 200 µg/kg		
Airtight coring devices	Refrigerated in coring device at 2 to 6°C	48 hours of collection[1]
	Frozen in coring device at −12°C	7 days of collection[2]
	Transferred at the laboratory into water preserved with NaHSO₄ to pH < 2 and refrigerated at 2 to 6°C	14 days of collection[2,3]
	Transferred into unpreserved water at the laboratory and refrigerated at 2 to 6°C	48 hours of collection[2]
Autosampler vials with water	Preserved with NaHSO₄ to pH < 2 and refrigerated at 2 to 6°C	14 days of collection[3]
	Unpreserved, refrigerated at 2 to 6°C	48 hours of collection[2,3]
Autosampler vials without water	Refrigerated at 2 to 6°C	48 hours of collection[2]
	Frozen at −12°C	7 days of collection[2]
High detection limits—VOC concentrations above 200 µg/kg		
Containers with methanol	Refrigerated at 2 to 6°C	14 days of collection[2]
		Up to two months[3]
		28 days of collection[4]

[1] EPA, 1996a.
[2] EPA, 1999e.
[3] ASTM, 1998.
[4] ADEC, 1999.

The decision on the proper sampling technique will be influenced by each of these factors, and the selection process may go through several field trials until the most appropriate sampling option is established. To assist the reader in the selection of the proper sampling method, Example 3.2 addresses their advantages and limitations.

3.5.5.1   Soil with low VOC concentrations

The definition of low-level contamination depends on the contaminant nature and the analytical method to be used. The concentrations of aromatic and halogenated hydrocarbons that are below 200 µg/kg are considered to be low because they fall within the calibration range of the laboratory instruments used for their analysis. A gasoline concentration in soil of 5000 µg/kg, high as it may seem, would also be within the instrument calibration range of most analytical laboratories.

Soil that is suspected of containing low VOC concentrations may be collected for analysis in three different ways:

1. In airtight coring devices

Example 3.2: Advantages and limitations of soil sampling options for VOC analysis

Common to all airtight coring device and autosampler vial sampling:
- The lowest PQLs are achieved.
- Collocated samples must be collected in duplicates or triplicates for backup analysis and laboratory QC.
- Additional sample volume must be collected for moisture and carbonate determination.
- Dilutions and reanalysis are not possible.
- Alkaline or calcareous soils cannot be preserved with sodium bisulfate.

Airtight coring devices
- May be used only for cohesive, uncemented soils.

Autosampler vials with water:
- Trips blanks must be prepared similarly.

Autosampler vials without water:
- May be used for any soil type.

Containers with methanol:
- May be used for any soil type.
- Sample compositing is possible.
- Additional sample volume must be collected for moisture determination.
- Extracts may be diluted and used for multiple analyses.
- Trips blanks must be prepared similarly.
- Special requirements for shipping of a flammable material must be observed.
- Extracts are disposed of as hazardous waste.

---

2. In preweighed autosampler vials with water, preserved with sodium bisulfate (NaHSO$_4$) or with unpreserved water
3. In preweighed autosampler vials without water

Sampling with airtight coring devices

Sampling into airtight coring devices works best for relatively cohesive yet uncemented soils. This technique is not applicable for soils, which contain gravel or rocks or are too hard (cemented) for the device barrel to be pushed into the soil. Soil, which is too sandy or is excessively wet will allow easy penetration with the barrel, however if it is not sufficiently consolidated, it may fall out of the barrel before it is capped.

To sample with airtight coring devices, we must have access to exposed soil, which may be the ground surface, the bottom and sidewalls of an excavation, the stockpile face or side, or soil in the excavator bucket. We may also apply this technique for sampling of subsurface soil brought to the ground level in a split spoon sampler.

To sample with the En Core$^{(TM)}$ coring device, we need the following supplies:

- En Core$^{(TM)}$ airtight sampling device or equivalent
- En Core$^{(TM)}$ T-handle sampling tool or equivalent

- Sample containers for dry weight determination (VOA vials or 2-ounce jars) with labels
- Stainless steel spatulas, disposable plastic scoops, or wooden tongue depressors
- Chain-of-Custody Forms
- Disposable sampling gloves
- Safety glasses and other appropriate personal protective equipment (PPE)
- Paper towels
- Pens
- Field logbook
- Sample packaging and shipping supplies per Chapter 3.4

To sample with the En Core™ sampling device, we follow these steps:

1. Wear disposable gloves, safety glasses, and other appropriate PPE while sampling.
2. Identify the sampling point in the field as required by the SAP.
3. With a scoop remove the upper 2–3 inches (in) of exposed soil.
4. Holding the En Core™ coring device with the T-handle up and the cartridge down, insert the sampling device into the freshly exposed soil. Look down the viewing hole to make sure the plunger o-ring is visible, indicating that the cartridge is full. If the o-ring is not visible, apply more pressure to fill up the cartridge. Withdraw the sampling device from soil, and with a spatula remove excess soil from the bottom of the barrel to make the soil surface flush with the walls of the barrel. Wipe the barrel with a paper towel to remove soil from the external surface.
5. Cap the cartridge while it is still on the T-handle. Push and twist cap until the grooves are seated over the ridge of the coring body.
6. Disconnect the capped sampler by pushing the locking lever down on T-handle, and twisting and pulling the sampler from T-handle. Lock the plunger by rotating plunger rod until the wings rest against the tabs. Fill out the label with the sample information on the original En Core™ sampler bag and place the cartridge into the bag.
7. Collect three En Core™ samples for each sampling point as close as possible to each other.
8. As required by the SAP, collect a field duplicate sample into three En Core™ samplers as unhomogenized split samples obtained as close as possible to the location of the primary sample.
9. As required by the SAP, fill two additional devices for MS/MSD analysis with soil obtained as unhomogenized split samples as close as possible to the location of the primary sample. Clearly identify these samples on the label as 'MS' and 'MSD.'
10. Store the En Core™ sampler bags in a cooler with ice.
11. Collect 10–20 g of soil into a clean, labelled VOA vial or a jar for dry weight and carbonate presence determination (optional).
12. Document the sampling points in the field logbook; fill out the COC Form.
13. Pack samples for shipment as described in Chapter 3.4.

14. Ship the samples to the laboratory at soon as possible and by the fastest available delivery service.

The laboratory has several preservation options for samples in airtight coring devices, as specified in Table 3.2, after they have been received at the laboratory:

Option 1. Store in the airtight coring devices at 2–6°C; analyze within 48 hours of collection. Prepare for analysis as described in Option 4.

Option 2. Freeze in the coring devices at −12°C; analyze within 7 days of collection. Prepare for analysis as described in Option 4.

Option 3. Extrude the soil from the airtight coring devices into pre-weighed, labelled autosampler vials with PTFE-lined septum caps; a magnetic stir bar; and 5 ml of analyte-free water, preserved with sodium bisulfate to pH<2. Weigh to determine the sample weight. This method is appropriate for soils that will not effervesce in a reaction with the acidic water solution. A sample prepared in this manner is ready for analysis. Store at 2–6°C, analyze within 14 days of collection.

Option 4. Extrude the soil from the airtight coring devices into preweighed, labelled autosampler vials with PTFE-lined septum caps; a magnetic stir bar; and 5 ml of analyte-free water. Weigh to determine the sample weight. A sample prepared in this manner is also ready for analysis. Store at 2–6°C; analyze within 48 hours of collection.

Option 5. Preserve two soil samples per either one of the four above-described options; transfer the third sample into a vial with 10 ml of purge-and-trap grade methanol; store at 2–6°C; analyze methanol-preserved samples within 14 days of collection.

Preservation with an acidic sodium bisulfate solution cannot be used for soils that are calcareous (contain carbonates) or are alkaline. The resulting carbon dioxide gas release (effervescence) will strip volatile compounds out of the soil or create the gas pressure capable of shattering the glass vial. If the soil is suspected of containing carbonates, the laboratory must perform a preliminary effervescence test on a separate aliquot of each sample before preserving them with sodium bisulfate. Chemical changes in the sodium bisulfate solution may also take place, for example, the formation of acetone (Hewitt, 1999a; Zimmerman, 2000) or loss of styrene in storage, possibly due to polymerization (Hewitt, 1995a; Zimmerman, 2000).

The laboratory will introduce the required internal and surrogate standards through the septum with a small gauge syringe and then place the vials on the autosampler for analysis. Matrix spikes will be prepared in a similar manner.

Because each vial with the soil/water slurry can be analyzed only once, sample dilutions are not possible. To overcome this problem, we should always have a sample aliquot preserved in methanol. Methanol extracts are diluted in water for analysis, and because the laboratory needs very small amounts (5–300 microliters [μl]) of methanol extract, multiple dilutions and reanalysis of sample preserved in methanol can be easily performed.

Practical Tips: Sampling with airtight coring devices

1. Gather as much information as possible on the type and grain size of soil before choosing the airtight coring device sampling option. Nothing is more upsetting than getting to the site with expensive sampling devices and finding out that they cannot be used because the soil is too hard, too rocky, or too sandy.

2. When planning on the sodium bisulfate preservation, verify that the soil is not calcareous or alkaline. Supply a sample to the laboratory for compatibility testing, if this information is not available.

3. Get spare airtight coring devices for backup in the field.

4. If soil is too loose or wet to stay in the airtight coring device, use a spoon or a spatula to keep it in place until you cap the device.

5. Due to short holding times, keep in constant communication with the laboratory regarding the date and time of sample delivery.

6. Plan to spend approximately 25 percent more on VOC analysis when using airtight coring devices.

7. When sampling for VOCs, do not use solvent or aerosol lubricants for decontamination and repair of the drilling equipment: these materials are volatile and may contaminate the samples.

8. When sampling with airtight coring devices without methanol preservation backup, anticipate the possibility of deficient analytical results because sample dilution is not possible.

Sampling into preweighed vials

Sampling soil directly into autosampler vials enables us to preserve the samples and to prepare them for analysis immediately after collection. When considering this type of sampling, we should also evaluate the type and grain size of the soil and the possible presence of carbonates.

The vials, which are either prepared by the laboratory or obtained from a supplier, are certified precleaned autosampler vials with PTFE-lined septum caps and magnetic stir bars inside. The vials may contain 5 ml of analyte-free water or be empty. The vials with water may be unpreserved or be preserved with approximately 1 g of sodium bisulfate to pH < 2. (We need to select the type of the preservation, which is appropriate for the soil to be sampled.) The laboratory or the manufacturer will weigh the prepared vials with labels attached, record the weight on the label, and ship them to the field sampling team.

To sample into preweighed vials, we need the following field supplies and equipment:

● Sample containers: certified *precleaned*, *preweighed*, *labelled* autosampler vials with PTFE-lined septum caps and magnetic stir bars

- Sample containers for dry weight determination (VOA vials or 2-ounce jars) with labels
- Disposable 10 ml plastic syringes with barrels smaller than the sample vial neck and the needle end of the barrel cut off or commercially available disposable *non-hermetic* soil sampling syringe devices
- Stainless steel spatulas, disposable plastic scoops, or wooden tongue depressors
- Trip blank: a vial with 5 ml of water, preserved or unpreserved
- Resealing plastic bags
- Small hacksaw or a razor knife
- Chain-of-Custody Forms
- Disposable sampling gloves
- Safety glasses and other appropriate PPE
- Paper towels
- Pens
- Field logbook
- Sample packaging and shipping supplies per Chapter 3.4

To make 10 ml plastic syringes into coring devices, we evenly cut off the needle end of the barrel with a hacksaw or a razor knife. To sample soil with these devices or with commercially non-hermetic available soil sampling syringe devices, we follow these steps:

1. Wear disposable gloves, safety glasses, and other appropriate PPE while sampling.
2. Identify the sampling point in the field as required by the SAP.
3. With a scoop or a spatula remove the upper 2–3 inches of the surface soil.
4. Press a cut-off syringe with the plunger fully inside the barrel into freshly exposed surface soil. As you press the barrel into the soil, the plunger will go up. Collect the soil volume of 2–3 ml (approximately 5 g), remove the syringe from the soil surface.
5. Clean the outside surface of the barrel with a paper towel. By depressing the plunger quickly, extrude the soil into a preweighed sample vial. Clean the threads of the vials of soil particles with a paper towel and cap the vial with its own PTFE-lined septum cap.
6. Record the sample ID on the vial label, place the vial in a resealing bag, and put it on ice.
7. Collect three samples in this manner for each location as close as possible to each other. Keep the three vials with the same samples in one resealing bag.
8. With the same syringe collect 10–20 g of soil into a clean, labelled VOA vial or a jar for dry weight and carbonate presence determination (optional).
9. Dispose of the used syringe into an appropriate receptacle.
10. As required by the SAP, collect a field duplicate sample with a new syringe into three vials as unhomogenized split samples obtained as close as possible to the location of the primary sample.
11. As required by the SAP, collect two additional samples for MS/MSD analysis as unhomogenized split samples obtained as close as possible to the location of the primary sample. Clearly identify these samples on the label as 'MS' and 'MSD.'

12. Document the sampling points in the field logbook; fill out the COC Form.
13. Pack samples for shipment as described in Chapter 3.4.
14. Ship the samples to the laboratory at soon as possible and by the fastest available delivery service.

If the soil is not conducive to sampling with airtight coring devices or cut-off syringes and low VOC concentrations are a matter of concern, we may revise the procedures and collect the needed 5 g of soil with a spatula followed by placing it into a vial. The key is to do it fast in order to minimize the soil exposure to air.

Once the laboratory has received the samples, they may be prepared for analysis and preserved in several different ways as follows:

Samples in water preserved with sodium bisulfate

- Weigh to determine the sample weight; store at 2–6°C; analyze within 14 days of collection.

Samples in water, unpreserved

- Weigh to determine the sample weight; store at 2–6°C; analyze within 48 hours of collection.

Samples without water:

- Weigh to determine the sample weight; add 5 ml of analyte-free water through the septum with a small gauge syringe; store at 2–6°C; analyze within 48 hours of collection.
- Weigh to determine the sample weight; freeze at −12°C; analyze within 7 days of collection.
- Preserve two soil samples as described above; transfer the third sample into a vial with 10 ml of purge-and-trap grade methanol; store at 2–6°C; analyze within 14 days of collection.

This type of sampling has the same limitations as sampling with airtight coring devices: it is not suitable for soils that contain carbonates, and samples cannot be diluted or reanalyzed, unless one aliquot out of three is preserved in methanol. Studies have shown that frozen storage without the addition of water yields the most consistent results (Zimmerman, 2000).

3.5.5.2 Soil with high VOC concentrations

Soil samples with aromatic and halogenated hydrocarbon concentrations above 200 µg/kg or with gasoline concentrations above 5000 µg/kg are defined as samples with high VOC concentration levels. Because these concentrations exceed the calibration ranges of most analytical instruments, sample extracts must be diluted for proper quantitative analysis. EPA Method 5035 describes two sampling techniques for soil with high VOC concentrations:

Practical Tips: Sampling into preweighed vials

1. Prepare all of the needed cut-off syringes ahead of time. Making these syringes requires some practice.

2. Inexpensive disposable non-hermetic plastic soil sampling syringes are available from several manufacturers. These devices are not suitable for storing and shipping samples, only airtight coring devices are.

3. To minimize splashing, carefully and slowly extrude soil from the syringe into a vial with liquid.

4. Collect one sample at a time to prevent switching of preweighed vial caps or magnetic stir bars.

5. If the soil is not amenable for sampling with the cut-off syringes, use a spatula and transfer the sample into a vial as fast as possible.

6. Do not put any extra labels or clear tape on the vials; it will make an accurate sample weight determination impossible.

7. The trip blank should correspond to the mode of sampling.

8. Collect one sample aliquot into an empty vial, and request that the laboratory preserve the sample in methanol for high concentration VOC analysis.

---

1. Methanol preservation in the field
2. Conventional sampling in glass containers or core barrel liners

   Preserving soil with methanol is the most efficient way to arrest bacterial activity and to prevent VOC volatilization. Methanol is also a much more efficient extraction solvent than water, and VOC concentration data obtained from the methanol-preserved samples are higher in values than the data obtained from soil/water slurries of the same samples (Vitale, 1999).

Methanol preservation in the field
The analytical laboratory usually provides sample containers preserved with methanol; they are also available from several supply companies. These *labelled* VOA vials or 2-ounce glass jars with PTFE-lined caps contain 10 ml of purge-and-trap grade methanol and are preweighed by the laboratory or the manufacturer. The weight is recorded in the label. The containers are tightly sealed and shipped to the site in coolers with ice. To assure that there was no methanol leaks or evaporation due to imperfect cap seals, the containers are often reweighed at the site prior to sampling. To determine the feasibility and the practicality of a portable balance use, we need to evaluate such factors as the field conditions, the number of samples to be collected, and the duration of the sampling event.

   To sample soil into methanol-preserved containers, we need the following field supplies and equipment:

- Sample containers: *precleaned, preweighed, labelled* VOA vials or 2-ounce jars with 10 ml of purge-and-trap grade methanol
- Sample containers for dry weight determination (VOA vials or 2-ounce jars) with labels
- Container Tracking Log similar to one shown in Appendix 15
- Calibrated top-loading balance, preferably battery operated, with 0.01 g sensitivity
- 10 g and 100 g balance weights
- Balance brush
- Balance forceps
- Disposable 10 ml plastic syringes with barrels smaller than the sample container neck and the needle end of the barrel cut off or commercially available soil sampling syringe devices
- Stainless steel spatulas, disposable plastic scoops, or wooden tongue depressors
- Trip blank: a container with 10 ml of methanol
- Resealing plastic bags
- Small hacksaw or a razor knife
- Chain-of-Custody Forms
- Disposable sampling gloves
- Safety glasses and other appropriate PPE
- Paper towels
- Pens
- Field logbook
- Sample packaging and shipping supplies per Chapter 3.4

To sample into methanol-preserved containers, we follow these steps:

1. Place the top-loading balance on a clean even surface in a location sheltered from wind, dust, and rain. Ensure that the balance pan is level; center the level bubble inside the window, if necessary. Clean the pan with the brush.
2. Verify the balance calibration by placing a 10 g and a 100 g weight on the balance pan. Use forceps; take care not to touch the weights with your hands at any time.
3. Record the weight in the Container Tracking Log. The acceptable reading must be within ± 0.01 g. If this criterion is not met, obtain another balance prior to start of sampling.
4. Weigh all sealed containers with methanol and record the weight to the nearest 0.01 g.
5. Compare the obtained weight to the original weight of the container. The loss of 0.2 g or more indicates it has been leaking. Do not use this container for sampling, get a backup one.
6. If weighing the containers in the field is not a feasible option, start with Step 7.
7. Identify the sampling point in the field as required by to the SAP.
8. Wear disposable gloves, safety glasses, and other appropriate PPE while sampling.
9. Collect samples using the previously described procedure for sampling into preweighed vials. *One methanol-preserved container is sufficient for each sample.*

10. As required by the SAP, collect one container for a field duplicate and one container for MS and MSD analysis
11. After samples have been collected, weigh the containers again and record the weight on the Container Tracking Log. Send a copy of the log to the laboratory.
12. Document the sampling locations in the field logbook; fill out the COC Form.
13. Pack samples for shipment as described in Chapter 3.4.
14. Ship the samples to the laboratory as soon as possible and by the fastest available delivery service.

One of the greatest advantages of sampling with methanol preservation is the availability of a compositing option. To prepare composite samples, we must increase the volume of methanol proportional to the number of grab samples that will constitute one composite sample. For example, for a four-point composite, we will need one preweighed labelled 2-ounce jar with 40 ml of methanol. We will collect four 5-g grab samples with separate syringes directly into this jar using the steps of the above-described procedure. We may also collect grab samples into individual methanol-preserved containers and have the extracts composited by the laboratory.

Because methanol is a flammable liquid, special requirements are applicable for shipping of methanol-preserved samples by air or land delivery services. The IATA and DOT publications describe the exemptions, the weight limitations, the labelling and shipping requirements for flammable materials. These organizations require accurate package labelling and precise shipping paperwork; incorrect packaging may cause delays in sample delivery to the laboratory.

The laboratory will store the containers with methanol-preserved soil at 2–6°C. The holding time for analysis is 14 days as defined by EPA Method 5035. ASTM Standard D 4547-98 indicates that methanol-preserved samples do not exhibit VOC losses for a period of time of up to two months (ASTM, 1998). The state of Alaska Department of Environmental Conservation (ADEC) Method AK101 for gasoline analysis of methanol-preserved soil samples specifies a holding time of 28 days (ADEC, 1999).

For soil with expected high VOC concentrations, sampling with methanol preservation is the best option for almost any soil type. The exceptions are gravel or gravely materials and certain types of clayey soils. Sampling of these matrices presents an interesting challenge. To overcome the problem of obtaining a representative sample of such soil, collecting a larger volume of soil into a larger volume of methanol seems to be the only sensible solution.

Conventional sampling

Conventional sampling into glass jars or core barrel liners is the least desirable option in sampling for VOC analysis. When soil is sampled into glass jars, the occurring aeration causes an immediate loss of VOCs, followed by losses in storage and during the preparation for analysis. These highly variable losses may be as high as 90–99 percent for some analytes (Hewitt, 1999a).

Practical Tips: Methanol preservation in the field

1. Use all of the practical tips offered for sampling into preweighed vials.

2. Request at least 5 percent more containers with methanol for use as blanks and as a backup in case of leakage.

3. Weighing samples in the field is cumbersome and time consuming. Do it only when practical.

4. Clearly indicate on the sample label that the preservation chemical is methanol or the laboratory might mistake it for water.

5. Be familiar with the requirements for shipping flammable material (methanol-preserved soil samples) by air and land.

Sampling of subsurface soil into core barrel liners has been widely used until EPA Method 5035 came into effect. In this sampling method, after a core barrel liner has been filled with soil, it is capped with PTFE sheets or aluminum foil and plastic caps. These types of closure do not prevent the VOC loss from soil, which for some compounds may exceed 97 percent after only 6 days of cold storage (Hewitt, 1999). If sampling and storage in glass jars or core barrel liners cannot be avoided, samples must be delivered to the laboratory and analyzed within the shortest time after collection.

3.5.6 Sampling for semivolatile organic and inorganic compounds

Sampling for SVOCs and inorganic compound analysis is a simple procedure compared to sampling for VOC analysis. Samples of surface soil are usually collected with disposable implements and placed in glass jars. Samples of subsurface soil are retrieved with hand or powers augers, with direct push techniques, or by placing boreholes. Soil is brought to the surface from discrete depths in liners placed inside the augers, direct push rods, or split spoon samplers used in borehole drilling with hollow stem augers.

To sample soil for SVOC and inorganic compound analysis, we need the following supplies and equipment:

- Sample containers for surface soil sampling: precleaned jars and lids
- Sample containers for subsurface soil sampling: stainless steel or brass core barrel liners 6 inches in length with plastic caps
- Sample labels
- PTFE sheets (if liners are used)
- Stainless steel spatulas, disposable plastic scoops or wooden tongue depressors
- Steel trowels or shovels
- Decontamination supplies per Chapter 3.8
- Chain-of-Custody Forms
- Disposable sampling gloves
- Safety glasses and other appropriate PPE

- Paper towels
- Pens
- Field logbook
- Sample packaging and shipping supplies per Chapter 3.4

To sample surface soil, we follow these steps:

1. Identify the sampling point in the field as required by the SAP.
2. Wear disposable gloves, safety glasses, and other appropriate PPE while sampling.
3. Use a decontaminated shovel or trowel to remove the turf at the proposed sampling location, if necessary.
4. Remove loose debris and exposed soil from the top 1–2 in.
5. Collect a sample for VOC analysis first, if required by the SAP.
6. Collect a soil sample into a glass jar using a disposable sampling scoop or a decontaminated stainless steel spatula. Remove stones, twigs, grass, and other extraneous material. Fill the jar to the top with firmly packed soil.
7. Tightly cap the jar; label it and place on ice.
8. As required by the SAP, collect a field duplicate sample as close as possible to the location of the primary sample.
9. As required by the SAP, collect an additional jar for MS/MSD analysis as close as possible to the location of the primary sample. Clearly identify this sample on the label as 'MS' and 'MSD.'
10. Document the sampling locations in the field logbook; fill out the COC Form.
11. Decontaminate non-disposable equipment between each sampling point as described in Chapter 3.8.
12. Pack samples for shipment as described in Chapter 3.4.
13. Ship the samples to the laboratory at soon as possible and by the fastest available delivery service.

To sample subsurface soil with a split-spoon sampler, we follow these steps:

1. Wear disposable gloves, safety glasses, and other appropriate PPE while decontaminating sampling supplies and while sampling.
2. Decontaminate the liners and the split spoon as described in Chapter 3.8.
3. Identify the sampling point in the field as required by the SAP.
4. Drill to the desired depth and retrieve the auger. Attach the split spoon sampler with 3 liners inside to the sampling rod and insert into the boring. Drive the split spoon sampler further down the borehole according to ASTM Method D1586.
5. Retrieve the split spoon sampler from the borehole and open it.
6. If sampling for VOC is required, use an airtight coring device and quickly collect samples from the ends of the middle liner. Cap the lower liner with PTFE sheets and plastic caps, label it, and place on ice.
7. If sampling VOCs is not required, select a liner with the most recovery or from a specific depth, cap and label it, and place on ice. Use only the middle or the lower liner.

8. Discard the soil in the upper liner containing sluff (cavings) as IDW.
9. Document the sampling locations in the field logbook; fill out the COC Form.
10. Decontaminate the split spoon and other non-disposable equipment between sampling points as described in Chapter 3.8.
11. Pack the samples for shipment as described in Chapter 3.4.
12. Ship the samples to the laboratory at soon as possible and by the fastest available delivery service.

For this type of sampling, the collection of additional sample volumes for MS/MSD analysis is not possible. The liners, however, usually contain enough soil for the analysis of the sample itself and for MS/MSD analysis. In this case, we designate one of the field samples as 'MS/MSD' on the COC form. Collocated field duplicates are usually collected as the middle and the lower liners from the same depth interval or occasionally by placing another borehole next to the location of the primary sampling point.

Because different analyses need a different weight of soil as shown in Appendix 12, one sample container may be sufficient for all of the required analysis. The SAP should specify how many sample containers should be collected as illustrated in Example 3.3. Collecting excessive soil volumes increases the cost of sampling and shipping.

Example 3.3:   Combine soil for analysis and ship fewer containers

The SAP should contain a table that specifies the exact requirements for sample containers, preservation, and holding time as shown below.

Analysis	SW-846 Method	Container	Preser-vation	Holding time
TPH as gasoline	5035/8015B	3 En Core® devices	2–6°C	48 hours
VOCs	5035/8260B	3 En Core® devices	2–6°C	48 hours
TPH as diesel and motor oil	EPA 8015B	One brass liner	2–6°C	14 days for extraction; 40 days for analysis
Metals	EPA 6010B/ 7000			180 days for all metals, except mercury; 28 days for mercury
SVOCs	EPA 8270C			14 days for extraction; 40 days for analysis

3.5.7 Homogenization and compositing

Homogenization is a technique designed to enhance the sampling point representativeness. We use homogenization to make certain that a sample truly represents the properties of soil at a given sampling point and that a subsample taken from a sample container for analysis represents the true characteristics of the soil in the container.

Sample compositing is a technique that enables us to obtain a mean concentration of a group of samples at a reduced analytical cost. Composite samples are several grab samples collected at different times or in different sampling points and mixed together to form a single sample. Equal volumes or weights of grab samples are mixed together to characterize the average contaminant concentrations.

We composite or homogenize samples using a procedure, which is often referred to as the *four quarters* or the **quartering** method. The required equipment consists of a stainless steel pan or a tray and a stainless steel spoon or a scoop. To eliminate the need for decontamination between samples, these inexpensive supplies may be discarded after each sample.

To perform the *four quarters* method, we follow these steps:

1. Obtain equal volumes of grab samples to be composited or 3–6 volumes of a sample to be homogenized.
2. Place the soil into a mixing tray; divide the soil into four quarters; and thoroughly mix each quarter separately with a spoon.
3. Mix the four quarters together in the center of the tray.
4. Divide the soil into four quarters again; mix each quarter separately; combine them in the center of the tray.
5. Repeat the procedure as necessary to achieve complete homogenization as evidenced by the physical appearance of the soil.
6. Once mixing is completed, divide the soil in the tray into two halves.
7. Fill a sample container by alternating the soil from each half.
8. Seal and label the container.

Homogenized split QA and QC samples are prepared as follows:

1. Collect two volumes of soil from the same sampling point.
2. Transfer the soil from sample containers into a mixing bowl.
3. Mix it thoroughly using the four quarters method.
4. Divide the soil into two halves and fill two sample containers by alternately placing the soil from each half into each sample container.
5. Seal and label the containers.

Practical Tips: Homogenization and compositing

1. Never composite or homogenize samples for volatile contaminant analyses (solvents, gasoline, jet and light diesel fuels, mineral spirits.)

2. When compositing in the field, to avoid using a balance, composite equal volumes of soil. Laboratories, however, prepare composites based on equal weight of soil subsamples.

---

continues

Practical Tips: Homogenization and compositing (continued)

3. The higher the moisture and clay contents in soil, the harder it is to composite or to homogenize it.

4. To avoid the decontamination of mixing equipment and equipment blank analysis, use disposable mixing tools whenever possible. Disposable equipment will be less expensive than the cost of analysis and the disposal of decontamination wastewater.

5. When stainless steel utensils are not available, aluminium foil, plastic sheeting, and plastic scoops are a viable alternative.

3.6   WATER SAMPLING

Sampling of water from surface sources, such as lakes, rivers, lagoons, ponds, or from groundwater aquifers presents its own universe of practices and procedures. Water samples may also be collected to represent rain, fog, snow, ice, dew, vapor, and steam. Because of a great variability of water sources and its physical states, there are a great variety of methods and tools for sampling water. In this chapter, we will focus on the basic procedures for sampling of groundwater, surface water, and tap water.

When sampling water from any source, the main issue is the one of sample representativeness. Proper sampling procedures will enable us to minimize variability at the sampling point and to maximize the sampling point representativeness, whereas the use of appropriate sampling tools and preservation techniques will ensure the 'collected sample' representativeness. The sampling frequency, the selection of proper tools and the types of QC samples will depend on the project DQOs. Other important variables, such as seasonal variations in contaminant concentrations or contaminant stratification, should always be taken into account while developing a sampling design.

Water samples are usually collected as grab samples. For regulatory compliance, as part of NPDES permits, composite samples may be required. Composite samples of water may be prepared either on a flow-proportional basis or as time composites. *Water samples for VOC, SVOC, TPH and oil and grease analysis are never composited.*

3.6.1   Reactions in water samples

Common to all water sampling procedures are several underlying issues related to the chemical reactions, which take place in the water samples between the time of collection and the time of analysis. Understanding the chemical processes that affect contaminants in a water sample is critical for selecting appropriate sampling tools and effective preservation techniques and in evaluating data quality.

When exposed to atmospheric conditions, contaminants in water samples undergo irreversible changes under the effects of oxygen in air, light, and, for groundwater samples, the differences in temperature and pressure. Table 3.3 summarizes the physical and chemical changes that may take place in water samples and the common practices for minimizing these adverse changes.

The use of sampling equipment and containers that are made of non-reactive materials and proper preservation and storage are essential for minimizing these changes. As demonstrated in Table 3.3, there are good reasons that justify the

Table 3.3 Physical and chemical changes in water samples

Mechanism	Change	Loss minimization/ preservation technique
Volatilization	Loss of volatile organic compounds	Proper sampling techniques, headspace-free sample containers, cold storage
Bacterial degradation	Loss of chemical contaminants due to microbial activity	Acid preservation to pH < 2, cold storage
Chemical reactions	Formation of trihalomethanes in treated water due to the presence of free chlorine	Preservation with sodium thiosulfate to destroy excess chlorine; cold storage
Precipitation	Loss of metals due to formation of salts, oxides, and hydroxides with low solubility in water	Preservation with nitric acid to pH < 2, storage at room temperature
Adsorption	Loss of metals due to adsorption to glass surface	Use of plastic containers, preservation with nitric acid to pH < 2
	Loss of oily materials due to adsorption to plastic surface	Use of glass containers, preservation with sulfuric acid to pH < 2
Absorption of atmospheric gases	Chemical oxidation due to exposure to oxygen in air Change in pH and conductivity due to carbon dioxide absorption	Prompt and proper preservation Measurements of field parameters during sampling
Photooxidation	Chemical changes due to photochemical reactions	Protection from exposure to light, use of amber glass bottles
Diffusion	Introduction of contaminants from man-made materials, such as solvents from polyvinyl chloride (PVC) materials and PVC cement, plasticizers, and phthalates from polyethylene and polypropylene materials	Use of inert materials (PTFE, fiberglass-reinforced epoxy materials); steam-cleaning of groundwater well components prior to installation
Leaching	Introduction of contaminants from man-made materials, such as metals from stainless steel, silicon and boron from glass, lead and tin from solder, solvents from duct tape	Use of inert materials, pH control, minimization of contact time with these materials

selection of containers and preservation requirements for water samples listed in Appendix 13, and we must observe these requirements at all times when sampling water.

3.6.2 Groundwater sampling

The main source of the drinking water throughout the world is groundwater. In the USA, approximately 40 percent of drinking water is drawn from groundwater supplies (Drever, 1997). By sampling groundwater monitoring wells and analyzing the samples, we determine the water quality and make decisions on its suitability for beneficial uses.

As with soil sampling, the most important issue in groundwater sampling is sample representativeness or the accuracy with which a groundwater sample represents the formation water. The appropriate groundwater well design and use of non-contaminating construction materials are important factors that affect sample representativeness. Equally important is the reduction of sample variability achieved by the selection and implementation of proper sampling techniques. A combination of all three factors (correct well design, compatible construction materials, and appropriate sampling technique) will assure that a sample represents the true conditions in the aquifer.

The stagnant water in the well casing, which has been exposed to air inside the casing and has undergone changes due to oxidation and degassing, does not represent the formation water. To obtain a representative sample, prior to sampling we remove (purge) the stagnant water and allow for the screened interval of the well to refill with the fresh formation water or collect a sample directly from the screened interval with a minimum mixing of the fresh and the stagnant water.

A newly installed, i.e. a greatly stressed well, must stabilize for at least 24 hours after it has been developed in order to produce a representative sample; in certain hydrogeological settings, stabilization time for new wells may exceed a week. After the well fills with fresh formation water, samples must be retrieved with a minimum exposure to air, with the least amount of resuspended sediment, and in a manner that prevents degassing.

There are several groundwater sampling techniques in use today.

1. *Conventional—purging and sampling using a pump or a bailer*
   The conventional sampling method consists of purging a well with a high-rate pump or by bailing to remove a predetermined volume of water (usually 3–5 casing volumes) and allowing the well to stabilize using water quality parameters as well stabilization indicators. Samples are then brought to the surface with a pump or with a bailer. This method induces a significant stress in the water table due to the fast pumping rates and due to the insertion of bailers that may increase turbidity and aerate the water column. Other drawbacks include operator's technique variability; the large volumes of purged water incurring higher disposal costs; and inadequate field filtration procedures for dissolved metal analysis.

2. *Low-flow minimal drawdown or micro-purge technique*
   The low-flow minimal drawdown purging and sampling technique, also known as the micro-purge method, employs a low flow rate pump to collect a sample of formation water without evacuating the stagnant water from the well casing. Limited purging with a low flow rate pump is still necessary to eliminate the effects of mixing of the stagnant and the formation waters when the pump intake travels through the water column. After the water quality parameters have indicated that the well is stable, the same pump is used for sampling. The

relatively low purging and sampling pumping rates approximate the natural hydrological flow through the formation and cause none or minimal drawdown and stress of the water table. Although this technique may in some cases require higher initial capital costs, it offers several advantages compared to conventional sampling: minimal disturbance at the sampling point; less operator variability; reduced volumes of purged water for disposal; in-line filtering; shorter overall sampling time (Puls, 1995).

3. *Direct push techniques*
   Direct push techniques allow a rapid collection of groundwater samples because they are performed without the construction of permanent monitoring wells. Stabilization parameters are not monitored during direct push sampling. Groundwater samples from discrete depths are retrieved with either a Geoprobe® or a Hydropunch® sampler or with a similar device. Geoprobe® is a vehicle-mounted hydraulically powered percussion/probing apparatus that may be used for the collection of soil, groundwater, and soil vapor samples. Hydropunch® is a groundwater sampling device that can be driven into the subsurface by a Geoprobe® or a cone penetrometer truck. Samples collected with direct push techniques allow delineating groundwater plumes in a rapid manner, particularly when they are analyzed in real time at a field analytical laboratory.

4. *Passive Diffusion Bag Samplers*
   Passive Diffusion Bag (PDB) samplers are an innovative technology for groundwater sampling. These disposable samplers are sealed low density polyethylene (LDPE) tubes filled with deionized water and lowered into the screened interval of a groundwater monitoring well. The LDPE wall pore size is only about 10 angstroms. Organic compounds that are present in groundwater diffuse through the PDB wall, and eventually their concentrations inside the PDB and in the groundwater equilibrate. A PDB sampler stays inside the well for a length of time sufficient for the full equilibration to take place. The PDB is then retrieved from the well; the water is transferred into sample containers and analyzed. PDB samplers provide the data for a particular interval in the well and allow the characterization of a complete water column. They are inexpensive and easy to use and they reduce or completely eliminate purge water produced with conventional or micro-purge sampling techniques. PBD samplers may not be appropriate for all classes of organic compounds; their use has been proven mainly for VOCs. More information on PDB samplers may be found on the EPA Technology Innovation Office Web page at http://clu-in.com.

Although the first two techniques have the same basic steps (the well is purged, stabilized, and samples are collected), we implement them using different tools and procedures. Common to these techniques are the measurements of well stabilization parameters, which allow us to judge whether the water in a purged well represents the formation water.

3.6.2.1 Well stabilization parameters
After the well has been purged, fresh water from the aquifer refills the screened interval and enables us to collect a sample of the formation water. The process of

purging creates disturbances in the well and its vicinity: the mixing of stagnant and fresh water caused by placing of water level measurement or purging devices; the changes in the concentrations of atmospheric gases; and increased turbidity due to resuspension of sediment particles. After the purging is completed, we must verify that the effects of these disturbances have subsided, and that well has stabilized and may be sampled.

We use water quality indicators, such as pH, temperature, conductivity (specific conductance), dissolved oxygen, oxidation-reduction potential (ORP), and turbidity, as groundwater well stabilization parameters. Stable values of three consecutive measurements of these parameters are considered an indication of a stabilized well.

It is best to establish specific stabilization parameters and their performance criteria for each well based on such factors as the aquifer properties; the well construction detail; the nature of contaminants of concern; the sampling technique; and field instrument specifications. However, this may be achievable only for wells with historical data on a long-term monitoring program. As a general practice, we may use the guidelines for the stabilization parameter criteria presented in Table 3.4.

Table 3.4 lists the parameters in the order by which they would usually stabilize. Temperature and pH are usually the first ones to stabilize because they are not very sensitive to the influx of fresh water. That is why they are not as good stabilization indicators as the other parameters. Their measurements, however, are important for data interpretation and should always be made during groundwater sampling. Dissolved oxygen content and turbidity, which is the measure of the particulate matter content in water, are typically the last ones to stabilize.

For conventional sampling, the most commonly used parameters are the pH, temperature, and conductivity. The other parameters may take a long time to stabilize

Table 3.4 Groundwater stabilization parameter performance criteria

Stabilization parameter	Conventional purging and bailing	Low-flow micro-purge
pH	±0.2 pH units[1]	±0.2 pH units[1] ±0.1 pH units[2]
Temperature	±1°C[1] ±0.5°C[3]	±1°C[1] ±0.5°C[3]
Conductivity (specific conductance)	±10%[1]	±10%[1] ±3%[2]
Oxidation-reduction potential	Not used	±10 mV[2]
Dissolved oxygen	Not used	±10%[2]
Turbidity	Not used	±10%[2]

[1] DOE, 1996.
[2] Puls, 1996.
[3] USACE, 1994.
mV denotes millivolt.

in a disturbed well, and that is the reason why they are hardly ever used in conventional sampling unless there is a specific need for these measurements. Micro-purge sampling produces less disturbance in the well, and dissolved oxygen content, ORP, and turbidity are commonly measured and used as well stabilization indicators in this type of sampling.

As shown in Table 3.4, acceptance criteria for stabilization parameters may have a range of recommended values. Because some are more stringent than others, they should be used with caution and be adjusted as necessary based on the knowledge of site-specific conditions. For example, the criterion for turbidity of ± 10 percent may be too stringent for some hydrogeological conditions or for the conventional sampling method. On the other hand, this criterion may be easily achieved for wells with dedicated submersible pumps and non-turbid, clear groundwater.

Unnecessarily stringent or unrealistic well stabilization parameter acceptance criteria are the cause of excessive water removal during well purging, which may drive some wells to dryness. When preparing sampling and analysis plans for unknown sites, we can only arbitrarily select and propose the commonly used acceptance criteria and apply them during baseline sampling. Once initial information on stabilization parameter performance is obtained, acceptance criteria should be revised to reflect specific conditions in the well.

Generally, the indiscriminant application of prescriptive protocols to groundwater sampling is not practical because there are too many variables affecting the water quality. Groundwater sampling procedures should follow common industry practices and at the same time be site-specific or even specific to individual wells. A characteristic hydrological and geological setting at every separate site; the seasonal variations in depth to water; the variability in the annual precipitation; the location of each well within a site; and even the selection of sampling tools—all these factors make every groundwater sampling situation unique. That is why the input from an experienced hydrogeologist who is familiar with the conditions at a specific site is critical in developing and applying the proper procedure.

3.6.2.2 Filtration

Organic and inorganic contaminants have two mobility mechanisms in groundwater: transport by colloidal particles and as a dissolved matter. Colloids are fine particles of clay, oxides, organic materials, and microorganisms that have an ability to react with certain classes of contaminants. Because of their small size, 0.001 to 1 micron (μm), colloidal particles form a suspension in water, creating turbidity. Especially susceptible to colloidal transport are surface-reactive molecules and charged particles: radionuclides, cations of heavy metals (lead, copper), anionic inorganic contaminants (hexavalent chromium, arsenic), and high molecular weight organic compounds with low solubility in water (PAHs, PCBs). Low molecular weight molecules, for example VOCs, which have higher water solubilities, are transported in groundwater primarily in a dissolved phase.

In addition to colloidal particles, groundwater may contain coarser particles of clay, silt, and sand. These particles may precipitate in the sample container and form a visible layer of sediment at the bottom. The amount of sediment in a water sample often depends on the sampling technique and cannot be accurately reproduced. The sediments may contain substantial concentrations of contaminants (metals, high molecular weight organic compounds), and if analyzed as part of a water sample, will create a high bias in the obtained concentrations. On request, analytical laboratory may carefully decant the water from the sample container to separate it from the excess of sediment prior to the extraction for organic compound analysis. If water samples with sediment are preserved with acid for metal analysis, the sediment becomes fully or partially dissolved and elevates the metal concentrations in the sample.

As a rule, water samples collected for the determination of dissolved metal concentrations are filtered in the field through a 0.45 μm filter. Most of the regulatory programs recognize this procedure as the industry standard, which generates reproducible results. This practice, however, has been scrutinized as a source of errors in estimating the mobile or truly dissolved species (Puls, 1989). The decision on the groundwater filtering procedure depends on the DQOs, and the choice of the filter pore size depends on the intended use of the data. If the purpose of sampling is the evaluation of colloidal mobility, samples should not be filtered; however, all necessary measures should be taken to reduce the sediment matter in the sample (EPA, 1995a). If the purpose of sampling is to obtain the dissolved contaminant concentrations, we should filter the samples in the field through a 0.45 μm filter. To further reduce the influence of suspended colloids and to better approximate the true dissolved concentrations, we may use filters with pore sizes as small as 0.1 μm.

For dissolved metal analysis, field filtering is essential; for other analyses it may be a choice that we make based on the DQOs during the planning phase of the project. Samples for VOC or other purgeable constituent analyses are never filtered; neither are, as a general rule, samples collected for SVOC analyses. If the purpose of sampling is to determine the true dissolved SVOC concentrations in groundwater versus dissolved plus colloid-transported concentrations, filtration in the field should be considered.

If in the course of filtration the groundwater is exposed to air, it will undergo changes due to oxidation. That is why it is best to filter groundwater in the field with in-line disposable cartridge filters. In-line filters are particularly useful during sampling with the low-flow micro-purge method. During conventional sampling by purging and bailing, we may filter samples with a hand pump and a negative pressure filtration apparatus or use a pressurized disposable bailer system with a built-in filter. A drawback of the negative pressure filtration is sample aeration and the resulting oxidation. In-line filtering during low-flow sampling does not have this problem as the sampling pump delivers the sample under positive pressure.

Unfiltered and unpreserved samples are sometimes sent to the laboratory for filtering and preservation upon delivery. Since the delay, short as it may be, adversely affects the chemical and physical state of metals in groundwater, this practice should be avoided.

Practical Tips: Filtering in the field

1. When sampling for dissolved metals, filter the sample directly into a preserved sample container.

2. Make sure to get the in-line filters with the connector size matching the diameter of the pump tubing.

3. Carry a supply of spare filters for replacing the clogged ones.

4. Filter cartridges have a flow direction arrow. Make sure they are installed correctly.

5. Flush the installed filter prior to collecting a sample by passing the groundwater through it and only then collect the sample.

3.6.2.3   Conventional sampling

This chapter describes the conventional groundwater sampling procedure, which consists of purging the well with a high flow electrical pump, followed by retrieving samples with a bailer by hand. When purging a well by hand bailing, we will use the same basic procedure, although we will retrieve the volumes of purged water with a bailer instead of a pump.

Bailers are pipes with an open top and a check valve at the bottom, which are used to retrieve liquid samples from groundwater monitoring wells, tanks, pits, and drums. When a bailer is lowered on a cord into the water, the bottom check valve opens allowing the bailer to fill with water and to sink. As the filled bailer is lifted from the water, the check valve closes to retain the water. A sample is obtained from the bottom end of the bailer by opening the check valve with a bottom-emptying device.

Bailers are manufactured of inert materials, such as polyethylene, PVC, PTFE, or stainless steel, and have various construction designs, lengths, and diameters. Some bailers are weighted to facilitate easy sinking into the water. The most commonly used bailer for groundwater sampling is 3 ft long, with a 1.5 in internal diameter and a 1 liter capacity.

The most economic bailers are precleaned disposable ones made of polyethylene or PVC as they do not require decontamination. Non-disposable bailers that are used for sampling of multiple wells must be decontaminated between wells. On long-term monitoring programs, more expensive PTFE bailers are often dedicated to each well; these bailers may need decontamination prior to sampling.

Selecting equipment

Well purging is a procedure, which we use to remove stagnant and mixed water in the well in order to collect a sample representative of the formation groundwater. In conventional sampling, 3–5 well casing volumes are removed prior to sampling. Depending on the groundwater recharge rate in the well, some wells may go dry before this volume has been removed. If the well is purged dry, it may be sampled after it has recharged to approximately 80 percent of its volume.

Wells should be sampled immediately after purging or after they have recharged. The delay in sampling will result in a loss of VOCs from groundwater in the well and in oxidation changes due to contact with air above the screened interval. If possible, the period of time from the completion of purging to the start of sampling should be the same each time the well is sampled.

Wells may be purged by bailing or by using electrical pumps. Bailing by hand is not a desirable approach because it causes a great stress in the water table and is tedious and time consuming. Nevertheless, it is often done, particularly when single wells or a small number of wells are to be sampled and financial or logistical constrains do not allow obtaining electrical pumps. Using pumps is a much more efficient approach to purging. Table 3.5 compares advantages, disadvantages, and applicability of various

Table 3.5   Comparison of groundwater sampling apparatus performance

Power source	Effect on VOCs	Decontamination	Water volume	Groundwater depth
Bailer—not very effective in deep wells or wells with a lot of water				
Manual operation	Degassing	Use disposable bailers	One liter	No restriction
Submersible pump				
Electric	No degassing at low flow rates ranging from 100 to 300 ml/min	Difficult after highly contaminated samples	No limit	No restriction
Bladder pump—economical if dedicated to long-term monitoring				
Compressed air or gas	No degassing	Typically dedicated and do not require decontamination	No limit	No restriction
Hand pump—not very effective in deep wells or wells with a lot of water				
Manual operation	Degassing	Does not need decontamination	Small volume	No restriction
Centrifugal pump				
Electric	Used for purging wells only	Difficult after highly contaminated samples	No limit	Relatively shallow (to 75 ft)
Peristaltic pump				
Electric	No degassing at low flow rates ranging from 100 to 300 ml/min	Does not need decontamination	No limit	Shallow (to 25 ft)

groundwater pumps. Suction pumps, such as peristaltic and centrifugal ones, are effective only in relatively shallow wells with the depth to water of 25–75 ft. For deeper wells, we use positive displacement submersible centrifugal and piston-driven pumps or gas-operated bladder pumps.

Well volume calculations

To calculate the well casing volume we need to know the depth of the well, the diameter of the casing, and the depth to groundwater on the day of sampling. Having this information on hand, we can calculate the well volume, using the following formula:

Volume: $$V = H \times F$$ —H is the difference between depth of well and depth to water or height of water column
—F is the factor for one foot section of casing from Table 3.6

To record the well parameters and the well volume calculations and to document groundwater sampling, we use a Groundwater Sampling Form similar to one shown in Appendix 16.

Table 3.6   Volume of water in one-foot or one-meter section of well casing[1]

Diameter of casing		F Factor	
Inches	Centimeters	Gallons/feet	Liters/meter
2	5.1	0.16	2.03
4	10.2	0.65	8.11
6	15.2	1.47	18.24
8	20.3	2.61	32.43
10	25.4	4.08	50.67
12	30.5	5.88	72.96

[1] USACE, 1994.

Sampling supplies

Before a well is sampled, as a health and safety measure, we may record organic vapor contents in the wellhead with a field photoionization detector (PID). Chemical odors emanating from the well may overpower the sampler and warrant the use of appropriate PPE.

To sample with the conventional purging and bailing technique, we need the following supplies and equipment:

• Water level meter with a tape measure graduated to 0.01 ft
• One of the pumps listed in Table 3.5 that is appropriate for the well

- Power source (a battery, a generator, a stationary power supply source)
- Tubing appropriate for the selected pump
- A length of safety cable for submersible pumps
- Field portable PID
- Field portable temperature, conductivity, and pH meter with a flow-through cell and calibration standards
- Field portable dissolved oxygen and ORP probes (optional for conventional sampling)
- Turbidity meter with calibration standards (optional for conventional sampling)
- Bottom-filling, bottom-emptying bailers
- Nylon cord
- Hand reel or a portable winch
- Filtration apparatus and filters
- Plastic cups
- Stopwatch
- Preserved sample containers and labels
- Three plastic 5-gallon (gal) buckets
- Drums or other holding vessels for purged water
- Groundwater Sampling Forms
- Chain-of-Custody Forms
- Field logbook
- Pens
- Disposable sampling gloves
- Safety glasses and other appropriate PPE
- Paper towels
- Scissors
- Preservation chemicals and disposable pipettes, if preserved sample containers are not available
- Universal indicator paper strips for pH ranging from 0 to 14 and a reference color chart
- Sample packaging and shipping supplies per Chapter 3.4
- Decontamination supplies per Chapter 3.8

To purge a well with a pump and sample it with a bailer, we follow these steps:

Well volume measurements

1. Wear disposable gloves, safety glasses, and other appropriate PPE while sampling.
2. Confirm the well identification at the monitoring well riser.
3. Calibrate all field portable meters, probes, and the PID according to manufacturer's specifications. Record calibration information in the field logbook.
4. Open the well cap and measure organic vapor at the wellhead with a PID. Record the reading in the Groundwater Sampling Form.
5. Decontaminate water level meter as described in Chapter 3.8.

6. Slowly lower the meter probe into the well casing. The meter will make a 'buzzing' sound as the probe enters the water. Lift and adjust the probe till the sound fades out completely, and pinch the tape where it meets the top of the casing. For accurate measurements always use the permanent reference point on the well casing made by a surveyor during well installation. If there is no surveyor's reference point, for future reference identify your own with a permanent marker during the first sampling event. Record the depth to water to the nearest 0.01 ft or 1/8 in on the Groundwater Sampling Form.
7. Remove the water level meter from the well.
8. Calculate the well volume and the purge volume and record them on the Groundwater Sampling Form.

Well purging

9. Secure the decontaminated submersible pump with the safety cable and lower it to a depth between 1 ft from the bottom of the screened interval and the middle of the screened interval. The pump must not disturb the sediment at the bottom of the well.
10. Adjust the flow rate based on the volume to be purged, typically, 0.5 to 5 gallon per minute (gal/min). If the pumping rate is unknown, measure the evacuated volume in a graduated 5-gal bucket and monitor the time to establish the rate.
11. After one well volume has been removed, measure the indicator parameters (pH, temperature, and conductivity) either in a flow-through cell or after collecting a sample from the discharge end of the pump into a plastic cup. Record the measurements on the Groundwater Sampling Form.
12. Lower the pump as necessary to maintain the submergence in the water.
13. Continue pumping until indicator parameters are stable within the acceptance criteria referenced in the SAP for three consecutive measurements taken after one casing volume has been removed between readings. Record the measurements on the Groundwater Sampling Form.
14. Remove the pump from the well. (If *not* sampling with a bailer, go to Step 5 of low-flow micro-purge technique, described below in Chapter 3.6.2.4.)

Sampling with a bailer

15. Tie an unused or a decontaminated bailer to a nylon cord with a reliable knot. Using a hand reel or a winch, lower the bailer into the well and *slowly* let it submerge.
16. Lift the filled bailer from the well.
17. Slowly discharge the water from the bottom check valve into preserved sample containers. To minimize turbulence and exposure to air, let the water flow down the side of the sampling container.
18. First, fill the vials for VOC analysis by slowly discharging water from the bottom of the bailer until a meniscus forms above the neck of a VOA vial. Seal each vial with a PTFE-lined septum cap, invert the vial, and gently tap it. If an air bubble forms in the vial, discard the sample, and fill another container.

19. Wipe the VOA vials dry, label them, and place in resealing bags inside coolers with ice.
20. Fill sample containers for all remaining organic analysis, label them, and place in resealing bags inside coolers with ice.
21. Collect a sample for total metal analysis.
22. Assemble the filtration apparatus. Pour sample into the filter funnel and filter the water by creating a vacuum in the collection flask with a hand pump. If the filter becomes clogged with sediment, release the vacuum, replace the filter, and continue filtering. Transfer the filtrate into a preserved sample container.
23. Collect the samples for all remaining analysis as appropriate.
24. Wipe the containers dry, label and place them in resealing bags inside coolers with ice.

Post-sampling activities

25. Complete sampling records in the Groundwater Sampling Form and in the field log book; fill out the COC Form.
26. If necessary, decontaminate the pump and the bailer as described in Chapter 3.8.
27. Dispose of the purged water and decontamination wastewater into drums or other holding vessels.
28. Cover the well head; lock the well.
29. Pack the samples for shipment as described in Chapter 3.4.
30. Ship the samples to the laboratory as soon as possible and by the fastest available delivery service.

3.6.2.4   Low-flow micro-purge technique

The low-flow minimal drawdown or micro-purge procedure follows the same basic steps of the conventional sampling routine, including water level determination, well volume calculation, well purging, and sampling. We will need the same supplies and equipment, and to begin sampling, will follow the first eight steps of the procedure for conventional sampling described in Chapter 3.6.2.3.

Well volume measurements

1. Complete steps 1 through 8 of the Conventional Sampling Procedure described in Chapter 3.6.2.3. Leave the water level meter inside the well.

Well purging

2. Lower the intake of the pump and a dissolved oxygen meter probe into the well slightly above or in the middle of the screened interval. Purge the well at a rate 300–500 milliliter/minute (ml/min). Check the water level periodically to monitor the drawdown, which should not exceed 4–8 in. If the pumping rate is unknown, measure the evacuated volume in a graduated container and monitor the time to establish the rate.

3. Monitor indicator parameters (pH, temperature, conductivity, ORP) in a flow-through cell every 1–2 minute (min) during purging. At the same frequency, monitor dissolved oxygen and turbidity. Record all measurements in the Groundwater Sampling Form.
4. Continue purging until the well is stabilized.

Low-flow micro-purge sampling

5. Reduce the flow rate to 100 ml/min and start sampling from the discharge end of the pump. To minimize turbulence and exposure to air, let the water flow down the side of the sampling container. First, fill the vials for VOC analysis until a meniscus forms above the neck of a VOA vial. Seal each vial with a PTFE-lined septum cap, invert the vial, and gently tap it. If an air bubble forms in the vial, discard the sample, and fill another container.
6. Wipe VOA vials dry, label them, and place in resealing bags inside coolers with ice.
7. Increase the flow to 300 ml/min. Fill sample containers for all remaining organic analysis, label them, and place in resealing bags inside coolers with ice.
8. Collect a sample for total metal analysis.
9. To collect a sample for dissolved metal analysis, stop the pump, and insert a filter cartridge into the line. Restart the pump, flush it with a small volume of groundwater, collect a sample into a preserved container.
10. Wipe the container dry, label it, and place in a resealing bag inside a cooler with ice.
11. Collect the samples for all remaining analysis as appropriate.

Post-sampling activities

12. Complete steps 25 through 30 of the Conventional Sampling Procedure.

3.6.2.5 Field preservation
If samples are collected into unpreserved sample containers, we preserve them in the field immediately after collection using the following procedures.

Samples for VOC analysis

1. Fill a VOA vial with collected water to the top of the neck, but without a meniscus. This is a test vial, which we will use for establishing the volume of acid needed for sample preservation.
2. Counting the drops, add approximately 0.5 ml of 1:1 hydrochloric acid solution to the test vial; measure the water pH by dipping a pH strip into the vial and comparing it to the color scale.
3. If the pH is above 2, add a few more drops to the test vial until the pH value is below 2.

4. Fill three more vials and preserve them with the same number of acid drops prior to sealing them headspace-free with PTFE-lined septum lids.
5. Discard the test vial.

Samples for inorganic and metals analysis

1. Fill the sample containers.
2. Add several drops of the preservation chemical, shown in Appendix 13, to the container. Seal the container and invert it several times to mix the content.
3. Pour a small volume of the preserved sample into the container lid or into a paper cup and take a pH measurement by dipping a pH paper strip into the lid or cup and comparing it to the color scale.
4. Adjust the pH in the container as necessary.

Practical Tips: Preservation with chemicals in the field

1. If samples preserved with acid exhibit effervescence from carbonate decomposition, record this observation on the sample label and in the Groundwater Sampling Form.

2. If the effervescence is so strong that an air bubble forms in the VOA vial immediately after the sample has been collected, discard the sample and collect another one. Do not preserve the sample for VOC analysis; preserve all other containers as required.

3. VOA vials filled with water must be headspace free (without bubbles) at the time of collection. A small bubble can, however, form in a collected sample due to degassing of dissolved gasses while in cold storage.

4. If groundwater samples have a documented effervescence problem, request a 7-day holding time for volatile aromatic hydrocarbon analysis of unpreserved samples.

3.6.2.6   Summary of groundwater sampling

When studying the groundwater sampling procedures, one can get a wrong impression that groundwater sampling is a relatively fast, highly accurate, and meticulous procedure that is conducted in a virtually sterile manner. In fact, groundwater sampling is usually a lengthy, fairly inaccurate, dirty, tedious, and often frustrating process. It takes place in an environment that is far from being sterile or even marginally clean: in the middle of a field overgrown with weeds, on a busy city street, or in a crowded parking lot. Well risers are often rusty, filled with rainwater and insects. A peculiar feature of groundwater sampling is that we cannot see the sampled matrix and, instead of vision, rely upon tactile perception.

The inclement weather conditions contribute to the difficulties of groundwater sampling: sun, rain, snow, and wind interfere with the integrity of the sampling process and make fieldwork physically demanding. No matter how early in the morning we start out on our sampling trip, uncooperative wells and capricious field equipment make the workday extend beyond dark.

Groundwater refuses to submit to a regimented written procedure, and the sampling of it may be more of an art than science. The ability to bring groundwater to the surface and collect a representative sample depends on the experience, intuition, and individual sampling technique of the sampler.

Groundwater sampling always requires quick thinking on your feet, because field conditions may change from one sampling event to another and or throughout the day. When sampling groundwater, we often face variances from the SAP, especially when working with newly-developed wells. Our decisions related to water quality parameters, recharge rates, and other previously poorly known factors should be justified and documented and be later incorporated into the SAP to reflect the true field conditions.

For an inexperienced sampler, venturing alone into the field for a day of groundwater sampling is a daunting mission. Since so much in this process depends on the technique, training with an experienced sampler is the only method of learning this trade. A combination of proper technique and a good understanding of physical and chemical processes that take place in the well and in the sample is a key to successful groundwater sampling.

Practical Tips: Groundwater sampling

1. Change gloves between sampling points.

2. Have a good map that is drawn to scale and clearly identifies the well locations on the site.

3. Obtain the information on well construction detail; prepare sampling equipment that is appropriate for the wells.

4. Obtain the information on the well depths and screened intervals prior to sampling. If the total well depth is to be measured, do it after sampling has been completed.

5. Use plastic ties or PTFE tape to tie together equipment that enters the water column; never use duct tape inside the well.

6. Protect sample containers and field meters from contamination with dirt and dust by placing them into large shallow plastic pans or keep them covered in coolers or boxes.

7. Remember that the wellheads are locked; do not leave for the field without a key.

8. Always secure a submersible pump with safety cable before lowering it down the well.

9. To minimize the potential for cross-contamination from the pump, first collect samples from wells with the lowest expected contaminant concentrations.

10. For sampling of groundwater wells on a monitoring program, have a premeasured, dedicated sampling tubing for each well. Store them in separate, clearly labelled, clean resealing bags.

11. Obtain bailers that are precleaned, individually wrapped, and sturdy (non-bendable).

12. A fishing line is a good alternative to the nylon cord for lowering and lifting a bailer. Do not reuse the line, dispose of it together with the used bailer.

---

continues

Practical Tips: Groundwater sampling (continued)

13. Pour water into sample containers only from the bottom discharge end of the bailer.

14. The shiny side of the VOA vial septum is the PTFE side; it should always face down.

15. Always carry at least three 5-gal buckets for purged water collection; a tool box with basic tools (screwdrivers, wrenches, etc.); plenty of resealing bags of various sizes; twist-ties and plastic ties; PTFE tape; plastic pans; extra coolers; paper towels; pens with indelible ink; labels; garbage bags—all these things are vital in the field.

3.6.3   Surface water sampling

Surface water is defined as water that flows or rests on land and is open to the atmosphere. Lakes, ponds, lagoons, rivers, streams, ditches, and man-made impoundments are bodies of surface water. The following principles are common for all types of surface water sampling:

1. Start sampling in the areas suspected of being the least contaminated and finish in the areas suspected of being the most contaminated. Start sampling a stream farthest downstream and proceed upstream in order not to contaminate the downstream areas while disturbing the upstream water.
2. If a discharge point is present, start sampling at the discharge point, and proceed with sampling the receiving body of water on a radial sampling grid that starts at the discharge point.
3. Field parameter measurements (pH, temperature, conductivity, etc.) may be part of the sampling procedure as indicators of surface water quality (not stabilization indicators as in groundwater sampling).
4. Whenever possible, take field measurements directly in the body of water by lowering the field instrument probes into the water at the point where a sample will be taken.
5. Collect samples directly into the sample containers by lowering them under the water surface. Sample containers in this case should not contain preservatives. Preserve samples as appropriate after they have been collected.
6. When sampling directly into a sampling container, approach the sampling point from the downstream, place the container in the water with its neck facing upstream and away from yourself.
7. When sampling from a boat, collect samples upstream from the boat.
8. If a body of water has no current, entering it would create a possibility of introducing extraneous contamination. In this case, whenever possible, sample from the shore using a long-handled dipper.
9. When wading in the water, do not needlessly disturb the bottom sediment. Wait till the sediment settles prior to sampling.

The supplies and equipment for surface water sampling varies depending on the type of the body of water to be sampled and on the intended use of the data. If sampling personnel is planning on entering the water, personal protection, such as rubber boots or waders, will be needed; a boat or an inflatable raft and paddles may be needed for sampling deep surface water.

3.6.3.1   Shallow surface water sampling

To sample shallow water, we need following supplies and equipment:

- Rubber boots or waders
- Long-handled or telescopic handle dippers or ladles
- Field portable meters, probes, and calibration standards
- Sample containers and labels
- Nylon cord
- Preservation chemicals
- Disposable pipettes
- Universal indicator paper strips for pH ranging from 0 to 14 and a reference color chart
- Field logbook
- Chain-of-Custody Forms
- Disposable sampling gloves
- Safety glasses and other appropriate PPE
- Paper towels
- Pens
- Sample packaging and shipping supplies per Chapter 3.4
- Decontamination supplies per Chapter 3.8

To sample surface water directly into containers, we follow these steps:

1. Organize sampling supplies, preservation chemicals, and other equipment in a clean work area on the shore.
2. Calibrate field portable meters and probes according to manufacturer's specifications.
3. Wear rubber boots or waders, disposable gloves, safety glasses, and other appropriate PPE while sampling.
4. Locate the sampling point in the body of water.
5. Enter the water, approach the sampling point, and allow the sediment to settle or reach into the water from the shore.
6. Take the field measurements directly in the water at the sampling point and record them in the field logbook or collect a volume of water in an unpreserved container and measure the field parameters once in the work area ashore.
7. Lower the unpreserved sample containers one by one into the water without disturbing bottom sediment. Direct the necks upstream and away from yourself, if you are in the water. If sampling from a bridge, tie a length of the nylon cord around the container's neck and lower the container into the water. Fill containers to the neck and cap them.
8. Bring the containers to the work area.
9. Preserve the samples as described in Chapter 3.6.2.5.

10. Wipe the containers dry, label them, and place in resealing bags inside coolers with ice.
11. Complete field records in the field logbook; fill out the COC Form.
12. Pack the samples for shipment as described in Chapter 3.4.
13. Ship the samples to the laboratory at soon as possible and by the fastest available delivery service.

To sample surface water with a dipper or a ladle, we follow these steps:

1. Organize sampling supplies, preservation chemicals, and other equipment in a clean work area on the shore.
2. Calibrate field portable meters and probes according to manufacturer's specifications.
3. Wear disposable gloves, safety glasses, and other appropriate PPE while sampling.
4. Decontaminate a reusable dipper or a ladle as described in Chapter 3.8, or unwrap an unused disposable one.
5. Locate the sampling point in the body of water.
6. Lower the dipper into the water without disturbing the sediment.
7. Retrieve a volume of water, pour it into an unpreserved container, and measure the field parameters once in the work area.
8. Retrieve more water from the same sampling point; transfer the water from the dipper into preserved sample containers; cap the containers.
9. Bring the containers to the work area.
10. Wipe the containers dry, label them, and place in resealing bags inside coolers with ice.
11. Complete field records in the field logbook; fill out the COC Form.
12. Pack the samples for shipment as described in Chapter 3.4.
13. Ship the samples to the laboratory at soon as possible and by the fastest available delivery service.

3.6.3.2   Deep surface water sampling

Deep water may be sampled from a discrete depth or at several discrete intervals. When sampling from several discrete depths at one sampling point, the order of sampling is from the shallow to the deep intervals. The sampling point is approached from a boat or from the shore, a bridge, or a dam.

Deep water sampling procedures are similar to those for surface water sampling, the difference is in the sample delivery method. There are several types of discrete depth liquid samplers available today to perform this task, such as glass weighted bottles, Wheaton bottles, Kemmerer samplers, or electrical pumps.

The glass weighted bottles and Wheaton bottles have a common design feature: they stay unopened until they are lowered to a desired depth on a measured line. The bottle stopper is attached to a line and may be opened by sharply pulling the line once the selected depth has been reached. The liquid fills the bottle through the upper opening, and the bottle is quickly retrieved to the surface.

The Kemmerer sampler is a length of pipe with messenger-activated top and bottom stoppers. The liquid flows freely through the sampler, while it is being lowered to a desired depth on a measured line. Once this depth has been reached, a messenger is sent down the line to activate the top and bottom stopper closure. Because the Kemmerer sampler is typically made of brass, which is not an inert material, its use is limited.

Electrical pumps are a good alternative to traditional glass and brass samplers. They enable us to collect samples from a discrete depth directly into the sample containers using a measured length of the intake tubing. The appropriate pump type is selected based on the lift capacity and the analyte nature. Samples for VOC analysis may be collected with a pump at pump rate of 100 ml/min.

To sample with a discrete depth sampler, we need the following supplies and equipment:

- Boat or inflatable raft
- Discrete depth samplers or a pump
- Nylon cord or pump tubing
- Tape measure
- Field portable meters, probes, and calibration standards
- Preserved sample containers and labels
- Preservation chemicals
- Disposable pipettes
- Universal indicator paper strips for pH ranging from 0 to 14 and a reference color chart
- Field logbook
- Chain-of-Custody Forms
- Disposable sampling gloves
- Safety glasses and other appropriate PPE
- Paper towels
- Pens
- Sample packaging and shipping supplies per Chapter 3.4
- Decontamination supplies per Chapter 3.8

To sample surface water from a discrete depth interval, we follow these steps:

1. Organize sampling supplies, preservation chemicals, and other equipment in a clean work area on the shore.
2. Calibrate the field portable meters and probes according to manufacturer's specifications.
3. Wear disposable gloves, safety glasses and other appropriate PPE while sampling.
4. Decontaminate a reusable discrete depth sampler as described in Chapter 3.8.
5. Measure a length of nylon cord (or tubing, if a pump is used) equal to the discrete depth from which a sample will be collected and attach it to the sampling device.
6. Approach the sampling point either from the water or from a location on the shore.

7. If sampling from a boat, measure the field parameters directly in the water at the sampling point, if possible. Otherwise, collect an additional volume of water, transfer it into a spare unpreserved container, and measure the field parameters once in the work area.
8. Slowly lower the sampling device into the water to a desired depth without disturbing the sediment and agitating the water.
9. Collect a sample and retrieve the sampling device; bring it into the boat.
10. Transfer the water into preserved sample containers; cap the containers.
11. Bring the containers to the work area.
12. If unpreserved containers were used, preserve the samples as described in Chapter 3.6.2.5.
13. Wipe the containers dry, label them, and place in resealing bags inside coolers with ice.
14. Complete field records in the field logbook; fill out the COC Form.
15. Pack the samples for shipment as described in Chapter 3.4.
16. Ship the samples to the laboratory at soon as possible and by the fastest available delivery service.

3.6.4 Tap water sampling

To determine the drinking water quality or to verify effluent discharge compliance with permit requirements, we sample the water from a tap or a sampling port in a water delivery system. Analytical data for potable water and effluent discharge samples may be compared to concentration action levels, such as the drinking water MCLs or NPDES permit limitations. Yet again we should be concerned with the issue of sample representativeness, as a water sample collected from a tap must accurately represent the water in the pipes and in the other elements of a water distribution system.

The basic drinking water sampling procedure consists of flushing the stagnant water from the pipes followed by sample collection. To flush the pipes, we open the tap and let the water flow for a few minutes after which the samples are collected from the tap directly into sample containers.

In practice, a decision on the representativeness of the water flow is often established based on the sampler's knowledge of the system piping. A change in water temperature, however, is the most obvious indicator of a stable flow as the sampler can sense it even with a gloved hand. A stable temperature reflects a stable water flow.

Sampling from a port in a water treatment system does not require flushing due to dynamic water flow conditions in the pipes. The measurements of the effluent pH and temperature are often conducted during sampling of treated water, as these parameters are usually included in NPDES permits.

We should take precautions not to introduce contamination from a sampling port or tap into the sampled water. Prior to flushing, we may wash dirty taps and ports with a non-phosphate detergent and water to remove dirt and dust, and rinse them on the outside with the water from the tap. We should never sample out of hoses, sprayers, and other tap attachments and should always replace rusty or leaking taps prior to sampling.

Drinking water samples that are collected for microbiological analysis should not be contaminated with unrepresentative bacteria during sampling. The water flow should not be fluctuating, as sudden surges may dislodge sheets of bacterial growth within the system. The flow should be moderate, steady, and should not contain particles. The outside surface of the tap may be cleaned with bleach, and samples are always collected into sterilized containers.

On occasion, we need to collect static samples from pipe systems, water heaters, etc., to determine whether they contribute lead and other contaminants to the water. In this case, our goal is to collect the stagnant water that has been in a system for several days. Prior to static conditions sampling, the system should not be used for a period of time from three days to a week. Sampling may take place after a small volume of water in the pipe immediate to the tap has been flushed.

The protocols for sampling of industrial wastewater require that the pH of preserved samples be verified immediately upon sampling to assure proper preservation (APHA, 1998). The basis for this requirement is the fact that some of the industrial wastewater streams have a high basic pH or contain acid-reactive chemical components, which may neutralize the acids added to samples for preservation. For this reason, the pH of preserved wastewater samples is always measured during sampling and is adjusted, if necessary, to bring it into required range of values.

To sample water from a tap, we need to following supplies and equipment:

- Field portable temperature and pH meter and calibration standards
- Plastic cups
- Preserved sample containers and labels
- Preservation chemicals
- Disposable pipettes
- Universal indicator paper strips for pH ranging from 0 to 14 and a reference color chart
- Field logbook
- Chain-of-Custody Forms
- Disposable sampling gloves
- Safety glasses and other appropriate PPE
- Paper towels
- Pens
- Sample packaging and shipping supplies per Chapter 3.4

To sample from a tap or a port, we follow these steps:

1. Wear disposable gloves, safety glasses, and other appropriate PPE while sampling.
2. Calibrate the temperature and pH meter according to manufacturer's specifications.
3. Remove screen from the tap, if present.
4. If needed, wash the outside of the tap and rinse it with the tap water.
5. Open the tap and let the water flow.
6. If sampling for the NPDES permit compliance, collect a sample into a plastic cup and measure pH and temperature. Record the readings in the field logbook.

7. If the purpose of the sampling is to document the static conditions within the water system, fill the sample containers after a small volume of water has been flushed.

8. If sampling from a water treatment system port, open the port and immediately collect a sample directly into a sample container. If the purpose of the sampling is to document the dynamic conditions in the water system, let the water run 4–5 min.

9. Fill the sampling containers. Preserve the containers as described in Chapter 3.6.2.5, if unpreserved containers were used.

10. Wipe the containers dry, label them, and place in resealing bags inside coolers with ice.

11. Complete field records in the field logbook; fill out the COC Form.

12. Pack the samples for shipment as described in Chapter 3.4.

13. Ship the samples to the laboratory at soon as possible and by the fastest available delivery service.

Practical Tips: Tap water sampling

1. Change gloves between sample locations.

2. Never clean taps or sampling ports with chemicals (methanol, hexane, or alcohols) as they will contaminate the sampled water.

3. Cover sampling ports of outdoor treatment systems with plastic to protect them from dust and rain.

4. When sampling influent and effluent from the water treatment system, sample the effluent first and label all containers before sampling the influent.

5. When sampling water in a privately owned house, record the name, the mailing address, and the telephone number of the owner.

6. For microbiological analysis, collect samples of cold water; keep them refrigerated for no longer than 30 hours prior to analysis.

3.7 SURFACE SAMPLING WITH WIPES

We use wipe sampling to detect contaminants on *non-porous* surfaces, such as the surfaces of plastic or metal drums; transformer casings; various heavy equipment; walls; floors; ceilings; laboratory benches. Sampling with wipes allows transferring contaminants from a surface area of a known size onto the wipe material. The wipes are then analyzed, and the amounts of contaminants found on the wipe are related to the surface area.

Typically, we perform wipe sampling in the following situations:

- To document pre- and post-decontamination conditions of manmade structures (buildings, storage tanks, pipes)
- To characterize surface spills of unknown substances

• To monitor surface contaminants for health and safety purposes

Porous surfaces, such as wood, plaster, and concrete, are also often sampled with wipes. Surface sampling of porous materials, however, does not provide any information on the contaminants that may be trapped inside the pores, cracks, and seams of the sampled surface. A better way to characterize porous materials is to collect and analyze chip samples.

Common contaminants that are sampled with wipes include dioxins/furans, PCBs, pesticides, and other SVOCs, metals, and occasionally, VOCs. Obviously, we do not expect to detect any VOCs on surfaces exposed to air, however we often sample for them to satisfy a disposal or a health and safety requirement.

As with soil sampling, we may use probabilistic or judgment sampling strategies for wipe surface sampling. After we have defined the sampling strategy and selected the sampling points, we overlay a 100 square centimeter (cm^2) template (usually a 10 cm by 10 cm one) onto the sampling point, and wipe the area inside the template with a cotton or gauze swatch or a paper filter. The wipe is moistened in a chemical specifically selected for the target analytes. The type of wipe material used for sampling (gauze, cotton, filter paper) is usually inconsequential; the selection of the proper preservation chemical is very important. Depending on their chemical nature, different contaminants are sampled and preserved with different chemicals as shown in Table 3.7.

In wipe sampling, the same area cannot be sampled twice. To sample the same sampling point for a different class of contaminants, we delineate an *adjoining* area and wipe it with a chemical appropriate for the target analyte. To thoroughly cover the sampled area, the wiping motion inside the template is performed in two directions, horizontal and vertical, as shown in Figure 3.12. We place the used wipe into a preserved sample container and submit it to the laboratory for analysis.

We may composite the wipes from several sampling points for the same analysis by placing them into one sample container and submitting to the laboratory as a single sample. Field QC samples for surface sampling with wipes include collocated duplicates and trip blanks. To collect collocated field duplicates, we sample the adjoining areas with wipes moistened with the same chemical; place them in separate sample containers; and submit to the laboratory for the same analysis. Trip blanks for wipe samples are the sample containers with unused preserved wipes. Trip blanks are important for evaluating wipe data as they enable us to determine the background levels of contaminants in the materials that we use in sampling. Matrix spike samples

Table 3.7   Preservation chemicals for wipe samples

Target analyte	Chemical
SVOCs, dioxins/furans, PCBs, pesticides	Hexane
Metals, except hexavalent chromium	1:1 nitric acid
Hexavalent chromium	Deionized water
VOCs	Methanol
Cyanide	1% sodium hydroxide

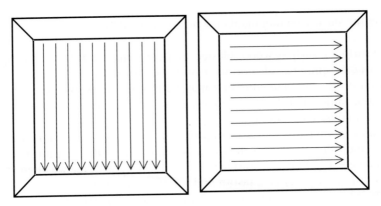

Figure 3.12   Patterns for wipe sampling on a template

are not collected for wipes; the laboratory determines analytical accuracy and precision using LCS/LCSD.

The laboratory will report results as milligram (mg) per wipe or microgram (μg) per wipe, and we can easily convert them into weight to surface concentration units of mg/100 cm² or μg/100 cm². These results cannot be converted into weight-to-weight concentration units (mg/kg or ug/kg), unless the thickness and the density of the wall that the surface represents is known.

To sample with wipes, we need the following equipment and supplies:

- Wipes: 11 cm glass microfiber Whatman filters; cotton or gauze squares or rectangles
- Stainless steel or PTFE-coated forceps
- PTFE squeeze bottles
- Preservation chemicals per Table 3.7
- Sample containers, preserved with 10–20 ml of chemicals, and labels
- Templates, 10 cm by 10 cm each
- Permanent marker
- Chain-of-Custody Forms
- Disposable sampling gloves
- Safety glasses and other appropriate PPE
- Paper towels
- Containers for waste chemicals
- Field log book
- Pens
- Sample packaging and shipping supplies per Chapter 3.4

To sample a surface area with wipes, we follow these steps:

1. Organize the sampling supplies in a clean area in the vicinity of the sampling point.

2. Wear disposable gloves, safety glasses, and other appropriate PPE while sampling.
3. Fill PTFE squeeze bottles with the chemicals and label the bottles with a permanent marker.
4. Using a squeeze bottle, rinse the forceps tips with a chemical to be used in sampling over the waste chemicals container.
5. Obtain a filter and fold it in half three times. It will look like a 45-degree segment of a circle. Insert the forceps 1 cm inside the folded filter and hold them tight to prevent the filter from unfolding. The tips of the forceps are inside the filter and are not exposed.
6. When using square or rectangular gauze or cotton wipes, fold them first in half lengthwise, then fold the edges to meet in the centre, and finally fold the wipe along the centre line. The edges will be inside the folded wipe. Insert the tips of the forceps inside the wipe to hold the edges. Two rectangular wiping surfaces will be available for use.
7. Slightly moisten the wipe with the appropriate chemical by spraying it from the squeeze bottle.
8. Holding the forceps with the wipe in one hand, overlay the template onto the area to be sampled.
9. Cover the surface inside the template with 10 overlapping passes of one side of the folded wipe, every time moving in the same vertical direction (upward or downward) as shown in Figure 3.12.
10. Turn the forceps over to switch to the unused side of the wipe and cover the surface with another 10 overlapping passes, every time moving in the same horizontal direction (for example, from left to right) as shown in Figure 3.12.
11. Place the wipe into a sample container with the appropriate chemical.
12. Seal the container with the lid, label it, and place it in a resealing bag inside a cooler with ice.
13. To collect another sample, repeat steps 5 through 12.
14. Complete sampling records in the field log book and fill out the COC Form.
15. Pack the samples for shipment as described in Chapter 3.4.
16. Ship the samples to the laboratory at soon as possible and by the fastest available delivery service.

Practical Tips: Surface sampling with wipes

1. Always include a wipe blank for each group of analytes when low concentrations are a matter of concern. For high contaminant concentrations, blanks may not be as important.

2. Use only high purity preservation chemicals obtained from the laboratory that will conduct analysis.

3. Do not use gauze or cotton wipes on rough surfaces because they tear easily.

---

continues

Practical Tips: **Surface sampling with wipes (continued)**

4. Make a stock of templates of plain paper or light cardboard.

5. Use the same template for collecting all samples at each sampling point; switch to a new template for each new sampling point.

6. Identify and mark the sampling points prior to start of sampling.

7. Document the sampling point condition (discoloration, rough surface, etc.) in the field logbook.

8. Areas greater or smaller than 100 cm^2 may be sampled; the wiped surface may be of any shape, as long as its area is known.

9. Use your judgment in estimating the sampling area for non-flat surfaces where templates cannot be overlaid, such as inside the pipes, ducts, or vents. In this case, collect the surface area samples without the use of a template.

10. For odd shaped areas, use templates of different dimensions, for example, 5 cm by 20 cm or of smaller sizes.

3.8 EQUIPMENT DECONTAMINATION

The use of non-disposable sampling equipment creates a potential for cross-contamination between samples. To eliminate the risk of cross-contamination, we must always decontaminate non-disposable sampling equipment and collect and analyze equipment blank samples to document the effectiveness of decontamination. Sampling containers that are not precleaned by the manufacturer must be also decontaminated prior to sampling.

Whenever possible, equipment decontamination, which is time-consuming and expensive, should be minimized. The best way to avoid decontamination is to use pre-cleaned, disposable equipment. This includes soil and water sample containers; pre-cleaned bailers and tubing for groundwater sampling; plastic or metal spoons, scoops, mixing bowls, and trays for soil sampling; ladles and other devices for surface water sampling. All of these sampling tools are available in a variety of disposable, inexpensive, single-use kinds. Dedicated pumps permanently installed in groundwater wells for a long-term monitoring program and dedicated tubing constitute another group of equipment that does not require decontamination.

Whenever we collect samples with reusable sampling equipment, we must decontaminate it between samples. The following types of sampling equipment require decontamination:

- Submersible pumps, water level meters
- Non-disposable dedicated bailers
- Large-scale soil sampling equipment (augers, corers, split spoons, and core barrel samplers)
- Non-disposable small sampling tools (bowls, spoons, etc.)

Decontamination procedure consists of washing the equipment with a detergent solution and rinsing it with tap water and with analyte-free water. The water from the final rinse is collected into sample containers and as the equipment blank is analyzed for the contaminants of concern.

The issue of the final rinse requires a special discussion. Several existing decontamination protocols recommend that the final rinse be done with ASTM Type II water (DOE, 1996; AFCEE, 2001). Many practicing professionals use high performance liquid chromatography (HPLC)-grade water for the final rinse. HPLC-grade water and ASTM Type II water are available from the distributors of chemical products, but they are expensive and *not necessarily free of the contaminants of concern*. The specifications of the ASTM Type II water for chlorides, sodium, total silica, and total organic carbon are inconsistent and unattainable by routine laboratory analysis (Kassakhian, 1994). HPLC-grade water, by definition, is free of organic compounds, which have ultraviolet absorbance. *The use of ASTM Type II or HPLC-grade water for final rinse does not enhance the project data quality; however, it increases the project cost.*

The water used for the final rinse should be free of contaminants of concern. Therefore, depending on the analyte nature, it may be distilled, deionized (USACE, 1994), or confirmed by chemical analysis as **analyte-free**. If possible, we should obtain analyte-free reagent water from the analytical laboratory that will be conducting analysis.

If VOCs are not among the contaminants of concern, bottled drinking water or commercially available deionized water often serves as an equitable substitute for analyte-free reagent water. (Certain VOCs are present in drinking water as the artifacts of water disinfection process.) If VOCs are among the project contaminants of concern, commercially available distilled water may be used for a final rinse. When bottled water is not available, and water of *unknown quality* is used for equipment blank collection, a source blank of such water is analyzed. The only situation when a source blank may be needed is when the sampled medium is water and low contaminant concentrations of organic compounds and metals are a matter of concern.

Several existing protocols require a solvent (acetone, methanol, isopropanol) rinse as part of equipment decontamination for VOC sampling and 1–10 percent hydrochloric or nitric acid rinse for metal analysis sampling (DOE, 1996; USACE, 1994). These practices, successful as they may be in removing trace level contaminants, create more problems than they are worth. Organic solvents are absorbed by the polymer materials used in sampling equipment construction and appear as interferences in the VOC analysis. Acid destroys the metal surfaces of soil sampling equipment and induces corrosion. The use of solvents and acids is a safety issue and it also creates additional waste streams for disposal.

To decontaminate field equipment, we need the following supplies:

- Three 5-gal buckets
- Laboratory-grade phosphate-free detergent
- Stiff-bristle brushes
- Plastic sheeting

- Potable water
- Analyte-free water
- Plastic or PTFE squeeze bottles
- Plastic or PTFE funnels
- Clean storage container or aluminium foil
- Sample containers for equipment blank samples
- Drums or other holding vessels for decontamination wastewater
- Disposable sampling gloves
- Safety glasses and other appropriate PPE

Water Sampling Equipment

To decontaminate water sampling equipment (pumps, water level meter, bailers), we follow these steps:

1. Before washing pumps, disassemble them and wash only the parts that come in contact with the sampled water.
2. Turn off the water level meter and wash only the part of it that contacted the water in the well.
3. Organize a decontamination station on plastic sheeting.
4. Wear disposable gloves, safety glasses, and other appropriate PPE.
5. Fill the buckets two thirds full of potable water. Dilute the detergent in one of the buckets per manufacturer's instructions.
6. Pour analyte-free water into a squeeze bottle.
7. Place equipment to be cleaned into the bucket with the detergent solution. Scrub it thoroughly with a brush.
8. Remove equipment from the detergent solution and rinse it with potable water in the second bucket.
9. Rinse equipment with potable water in the third bucket. Remove equipment from the bucket.
10. Spray analyte-free water from the squeeze bottle over the decontaminated surface that comes in contact with samples. Discard the rinsewater.
11. To collect equipment blank, place a clean funnel into the sample container. Spray the cleaned surface with analyte-free water from the squeeze bottle and direct the water from the cleaned surface into the funnel. Alternatively, pour the water from the cleaned surface directly into the sample container. (This step requires substantial manual dexterity and may need two persons.)
12. Air-dry the cleaned equipment before use on the next sample. If equipment will not be immediately used, keep it in a clean storage container or wrap it in aluminium foil.
13. Contain the wastewater in a drum or a holding vessel for future disposal.

Soil Sampling Equipment

To decontaminate small-scale soil sampling equipment made of metal (scoops, spoons, bowls, trays, core liners), we follow the same basic steps. In this case, the collection of equipment blank is not necessary because these tools are made of impervious materials. A clean surface must not have an oily residue or soil particles, and should feel clean and smooth to touch.

To decontaminate soil sampling equipment, we follow these steps:

1. Organize a decontamination station on plastic sheeting.
2. Wear disposable gloves, safety glasses, and other appropriate PPE.
3. Fill the buckets two thirds full of potable water. Dilute the detergent in one of the buckets per manufacturer's instructions.
4. Pour analyte-free water into a squeeze bottle.
5. Remove gross contamination from soil sampling equipment by dry-brushing.
6. Place equipment to be cleaned into the bucket with the detergent solution. Scrub it thoroughly with the brush.
7. Remove equipment from the detergent solution and rinse it with potable water in the second bucket.
8. Rinse equipment with potable water in the third bucket.
9. Rinse equipment with analyte-free water by spraying from the squeeze bottle.
10. Air-dry the cleaned equipment before use on the next sample. If equipment will not be immediately used, keep it in a clean storage container or wrap it in aluminium foil.
11. Contain the wastewater in a drum or a holding vessel for future disposal.

Large-scale downhole and heavy equipment, such as augers, split spoon and core barrel samplers, backhoe and excavator buckets, is also decontaminated between samples. Downhole equipment is decontaminated with pressure washing followed by a potable water rinse. Backhoe and excavator buckets are usually dry-brushed to remove gross contamination.

Practical Tips: Equipment decontamination

1. Whenever possible, use precleaned, disposable, or dedicated equipment to avoid decontamination

2. Spread plastic sheeting on the ground in the decontamination area to protect the ground surface from dirty water spills.

3. Do not attempt to decontaminate flexible plastic tubing: it has a strong tendency to retain organic compounds and cannot be effectively decontaminated.

4. Do not use solvent rinse on equipment for collection of samples for VOC analysis; solvents cause interference during analysis.

5. To avoid corrosion, do not use acid on metal equipment.

6. Use hot water if available; it is more effective in removing contaminants.

3.9 FIELD MEASUREMENTS

The measurements of water quality parameters (oxidation-reduction potential, pH, temperature, conductivity, dissolved oxygen, and turbidity) and the collection of field screening data with field portable instruments and test kits constitute a substantial portion of field work. Field measurements, such as pH, stand on their own as definitive data used for the calculations of solubility of chemical species and chemical equilibrium in water, whereas others serve as indicators of well stabilization or guide our decision-making in the field. Table 3.8 shows the diversity of field measurement

Table 3.8   Field measurement applications

Data use	Field measurement method	Laboratory analysis confirmation
Quantitative measurements of water quality parameters—temperature, pH, conductivity, oxidation-reduction potential, turbidity, dissolved oxygen		
• Groundwater stabilization parameters during sampling • Process monitoring	Field portable single or multiple parameter meters	Not possible; these are *in-situ* or short holding time measurements.
Semiquantitative field screening for organic vapor in soil headspace, at wellheads, at sampling ports of treatment systems		
• Site delineation • Excavation guidance • Excavated soil segregation • Groundwater sampling • Treatment system operation and monitoring	Organic vapor analyzers: photoionization and flame ionization detectors	Generally not necessary. Confirmation may be needed for compliance monitoring of air emissions.
Quantitative and semiquantitative screening for organic compounds in soil, water, and wipes		
• Site delineation • Excavation guidance • Excavated soil segregation • Surface cleanup verification	Immunoassay or colorimetric field kits	Establish initial correlation, confirm at a 10% rate for the duration of the project.
Quantitative or semiquantitative analysis for metals		
• Site delineation • Excavation guidance • Excavated soil segregation	Laboratory-grade or hand-held x-ray fluorescence analyzers	Establish initial correlation, confirm at a 10% rate for the duration of the project.
Water chemistry parameters—Hach Company quantitative or semiquantitative test kits and instruments		
• Treated water • Process water • Groundwater	Mostly colorimetric	Generally not necessary.

applications commonly used in environmental work and the types of data that may be collected for these applications.

There is no uniform operating procedure for field meters, field portable detectors, and field screening kits because different manufacturers make them in different formats. To use them correctly, we must strictly follow the manufacturer's instructions. When selecting a particular model, we need to evaluate its ruggedness, portability, selectivity, sensitivity, and reliability. To produce usable data, we must have a good understanding of the measurement mechanism, its applicability and limitations, and be concerned with the issues of field instrument calibration and maintenance. In this chapter, we will review some basic general chemistry definitions applicable to field measurements and focus on the common types of field analysis.

3.9.1 Water quality parameters

When groundwater is brought to the surface and stored in sample containers, the exposure to oxygen in air, the effects of daylight, and the differences in temperature and barometric pressure will cause irreversible chemical changes in a water sample, which in turn will change some of the water quality parameters. That is why water quality parameters should be measured *in situ* and with as little disturbance to the groundwater as possible.

The measurements may be obtained with single or multiple parameter meters that are available from several manufacturers (Horiba, Hydac, LaMotte, YSI, and others). Field meters are battery-operated and built into rugged, compact, often waterproof cases. The latest meter models are capable of measuring five or six parameters simultaneously with a submersible probe, eliminating the need of bringing the groundwater samples to the surface. Many models have automatic calibration features and perform multiple standard calibrations. Some meters store the readings in memory and may be interfaced with a printer. Although these advanced models offer the best efficiency and are easy to use, they are also expensive. Fortunately, less expensive and less sophisticated models will also suit most of our sampling needs.

If we do not own field meters, we will rent them for a sampling event. The rental company should calibrate the meters and make sure that they are in good repair.

Practical Tips: **General rules for field meter use**

1. Choose the field meter type and capabilities based on the parameter to be measured and the type of water to be sampled.

2. Always follow the manufacturer's instructions for the operation and maintenance of the field meters.

3. Monitor the calibration standards expiration dates and replace them as necessary.

4. Use submersible probes or flow-through cells for more accurate results and better efficiency in the field.

5. Do not wipe the probes—shake them to remove excess water and air-dry.

---

continues

Practical Tips: General rules for field meter use (continued)

6. Inspect probes for cracks, scratches, or salt crystallization: probes age with use and their performance deteriorates.

7. Store the probes per manufacturer's instructions, capped or in storage solutions.

8. Do not forget to remove the caps from the probe tips before making measurements.

9. Refill electrode probes with filling solutions per manufacture's instructions. Some inexpensive probes are not refillable, and should be replaced as their performance deteriorates.

10. Keep the meter cases dry, unless you are certain that they are waterproof.

3.9.1.1   Oxidation-reduction potential

Oxidation-reduction reactions (*redox* reactions) determine the chemical fate of many contaminants in groundwater and process water. Measuring the ORP (redox potential) enables us to evaluate the mobility and reactivity of non-metallic elements (sulfur, nitrogen, carbon) and metals in process water and to assess the types of redox reactions that take place in groundwater. We also use the ORP measurement as a well stabilization indicator in groundwater sampling.

Redox reactions involve a transfer of electrons and result in oxidation of one species and reduction of other. A reducing agent is a reactant that gives electrons and by doing so undergoes oxidation. An oxidizing agent is a reactant that accepts electrons and thus becomes reduced. The types of reducing and oxidizing agents that may be found in groundwater are illustrated in Example 3.4.

Example 3.4.   Reducing and oxidizing agents in groundwater

Reducing agent is the electron donor:
- Gives up electrons
- Contains the atom, which is oxidized
- The oxidized atom increases its oxidation state

Oxidizing agent is the electron acceptor:
- Receives electrons
- Contains the atom, which is reduced
- The reduced atom decreases its oxidation state

Reducing agents in groundwater include natural organic material, fuel hydrocarbons, chlorobenzenes, vinyl chloride. Oxidizing agents are dissolved oxygen, nitrate, Mn^{4+}, Fe^{3+}, sulfate, carbon dioxide. Polychlorinated benzenes and chlorinated solvents may become oxidizing agents under favorable conditions.

The ORP (p_e) can be expressed in units of electron activity as a negative logarithm of the electron activity (a_e) in a solution: $p_e = -log\ a_e$. The ORP can be also expressed in units of volts (E_h). As a matter of convention, the E_h measurements are referenced against a *standard hydrogen electrode (SHE)*.

ORP is measured as the electromotive force in an electrochemical cell, which consists of the reference electrode, a sensing electrode, and a potentiometer. Because the SHE is fragile and impractical, it is not used in field and laboratory ORP measuring probes; it has been replaced with silver/silver chloride or mercury/mercury chloride (calomel) reference electrodes. A typical reference electrode consists of a calomel or silver/silver chloride electrode, a salt-bridge electrolyte filling solution, and a so-called *liquid junction*. The liquid junction is a permeable interface in the reference electrode where the filling solution and the sample mix to complete the electrical circuit between the sample and the internal cell of the reference electrode. Because the filling solution slowly leaks out of the reference electrode, refilling the reference electrode is an important part of preventive maintenance.

The sensing electrodes are made of noble metals, usually platinum. The sensing electrode generates an electrical output based on electron transfer in the course of redox reactions, while the reference electrode generates a constant potential at a given temperature. The resulting potential is corrected for the difference between the potentials of the SHE and the reference electrode, making the measurement be relative to the SHE regardless of the reference electrode type. To signify this, a letter 'h' in the symbol for the redox potential E_h stands for hydrogen.

The accuracy of the ORP measurements depends on the temperature at which a measurement is taken. For solutions with reactions involving hydrogen and hydroxyl ions, the accuracy also depends on the pH of the water. In natural waters, many redox reactions occur simultaneously; each reaction has its own temperature correction depending on the number of electrons transferred. Because of this complexity, some of the field meters are not designed to perform automatic temperature compensation. The temperature correction for such meters may be done with a so-called *ZoBell's solution*. It is a solution of 3×10^{-3} mole (M) potassium ferrocyanide and 2×10^{-2} M potassium ferricyanide in a 0.1 M potassium chloride solution. The E_h variations of the ZoBell's solution with temperature are tabulated for reference, and the sample E_h is corrected as follows:

$$E_{h\ sample} = E_{h\ measured} + E_{h\ ZoBell\ reference} - E_{h\ ZoBell\ observed}$$

The ZoBell's solution and the tabulated reference values are usually part of the field meter kits. It may also be purchased separately from several field instrument manufacturers.

The ORP probes do not require calibration, however, we may use reference standards to verify the probe performance. Standards of a known potential are available from manufacturers; the ZoBell's solution may also serve a calibration check standard.

Field meters typically measure ORP and pH simultaneously with a probe that shares a common reference electrode for both measurements, but has two different sensing electrodes. The field ORP meters have a typical range of 0–1500 mV and an accuracy of 0.5 percent.

ORP levels in groundwater depend on the relative rates of the introduction of oxygen and its consumption by bacteria in the organic matter decomposition process. The E_h of a groundwater system will increase if the concentration of the oxidizing species increases relative to the concentration of the reducing species. Generally, the higher the dissolved oxygen content in groundwater, the higher the ORP. After all dissolved oxygen has been consumed, bacterial reduction of ferric (Fe^{3+}) to ferrous (Fe^{2+}) iron and sulfate (SO_4^{-2}) to hydrogen sulfide (H_2S) will typically take place, and the ORP will gradually decrease as an indicator of this process.

ORP may be accurately calculated based on activities of the reduced and oxidized species in water. Although field ORP measurements hardly ever correlate with the calculated values, they serve as reasonable estimates of the oxidation-reduction activity in natural waters.

3.9.1.2 Temperature and pH

The measurement of pH is one of the most important measurements in water chemistry. The value of pH defines the types and the rates of chemical reactions in water, and the fate and bioavailability of the living organisms. Together with temperature measurements, pH values serve as groundwater well stabilization indicators.

pH is a measure of the hydrogen ion activity (a_{H+}) in a solution, expressed as its negative logarithm: pH=−log a_{H+}. By definition, pH values range from zero to 14. Natural waters have a range of pH values of 4–9, and most are slightly basic because of the presence of naturally occurring carbonates and bicarbonates (APHA, 1998).

A pH meter consists of a potentiometer, a measuring probe, and a temperature-measuring and compensating device. The measuring probe is a combination electrode that houses in one body a sensing electrode and a reference electrode, which is submersed in an internal filling solution. The same reference electrode is used for the ORP and pH measurements, but the sensing electrodes are different. The electrode pair measures the electromotive force as a function of the hydrogen ion activity. The sensing electrode's output voltage varies with the hydrogen ion activity in the solution, whereas the reference electrode has a constant output.

The relationship between the electromotive force and pH is defined as a straight line. To establish the slope and the intercept of a meter system at a given temperature, the meter must be calibrated with standard solutions prior to use. A typical field meter kit includes three calibration standards, which are buffer solutions with known pH values (usually pH 4.01, 7.00, and 10.01 at 25°C). By immersing the probe into the buffer solution with pH 7, we establish the intercept (also called *offset* or *zero*) of the probe. If the reading is different from 7.0 at this point, we must adjust it with a control knob labelled 'Offset' or 'Zero.' The buffer solutions with the pH values of 4.01 and 10.01 allow us to verify and adjust the *slope (span)* of the calibration line.

After a calibration has been established, we may proceed with the sample measurements, which can be done with a submersible probe, in a flow-through cell, or in a plastic cup filled sampled water. Temperature variations are a source of several errors that affect pH measurements, causing changes in the equilibrium of the calibration solutions and samples and the drift of reference electrode. That is why field pH meters have a built-in temperature probe and a temperature compensation capability. Although the meter measures the electromotive force or potential in millivolts, the meter software delivers the readings in pH units. For routine work, we

organic matter decomposition, naturally occurring gases, and volatilized contaminants of concern. That is why these field analyzers produce the data that are mostly qualitative. Attempts to correlate field PID or FID soil vapor data with results of soil analysis have a very low rate of success. It is a well-established fact that neither PID nor FID provides meaningful screening data for samples with unknown contamination.

Field PIDs and FIDs are useful for process monitoring, where the contaminants are well defined by definitive methods of analysis. In this case, the screening results may be used for comparative purposes as indicators of a decrease or an increase in contaminant concentrations. An example of a meaningful PID application is the detection of leaks from the pipelines or the VOC measurements at the wellheads during operation of soil vapor extraction or other treatment systems. At hazardous waste sites, the PID serves as a useful screening tool for detecting the presence of VOC vapor in the wellheads of groundwater monitoring wells and in the headspace of drums and tanks.

3.9.3 Field screening methods

As part of SW-846, the EPA has validated and approved many immunoassay and colorimetric screening methods for a wide range of contaminants, such as petroleum fuels, pesticides, herbicides, PCBs, and explosives. Immunoassay technology uses the property of antibodies to bind to specific classes of environmental pollutants allowing fast and sensitive semiquantitative or qualitative detection. Colorimetric kits are based on the use of chemical reactions that indicate the presence of target analytes by a change in color. Table 3.9 presents a summary of EPA-approved screening methods and their detection capabilities.

In addition to test kits used in EPA-approved screening methods, a variety of other test kits are available from several manufacturers, for example, immunoassay test kits for BTEX in soil and water and for chlorinated solvents in water; colorimetric kits for the detection of lead; kits for a wide range of water quality parameter manufactured by Hach Company.

While field screening usually appears to be an excellent way to save time and money, we must be cautious in its application. In particular, the following questions must be answered before a screening method is selected for field use:

- How does the screening method work?
- How will the data be used?
- What is the detection limit?
- What are the limitations of the screening method (false negative and false positive results)?
- What are the cost savings?

In answering these questions, we must keep in mind that field screening methods produce mostly *semiquantitative* data that are not acceptable for confirmation sampling or site closure and that the screening method detection limits may not be sufficiently low for comparisons to the action levels.

Table 3.9   EPA-approved field screening methods

EPA screening method	EPA confirmation method	Range of detected concentrations or detection limit
4010—Screening for pentachlorophenol by immunoassay	8270, 8151, 8141	Soil: 0.1 to 10 mg/kg Water: 0.6 to 10 µg/l
4015—Screening for 2,4-D by immunoassay	8151	Soil: greater or less than 0.2, 1.0, and 10 mg/kg
4020—Screening for polychlorinated biphenyls by immunoassay	8082	Soil: 0.5 to 50 mg/kg Water: 0.5 to 10 µg/l Wipe: 5 to 250 µg/wipe Oil: greater or less than 5 ppm
4030—Soil screening for petroleum hydrocarbons by immunoassay	8015	TPH as gasoline: 10 to 10,000 mg/kg TPH as diesel: 15 to 10,000 mg/kg
4035—Soil screening for polynuclear aromatic hydrocarbons by immunoassay	8310, 8270	PAH as total: 1 to 1000 mg/kg PAH as phenanthrene: 0.2 mg/kg PAH as benzo(a)pyrene: 0.01 mg/kg
4040—Soil screening for toxaphene by immunoassay	8081	Greater or less than 0.5, 2.0, and 10 mg/kg
4041—Soil screening for chlordane by immunoassay	8081	Greater or less than 0.02, 0.1, and 0.6 mg/kg
4042—Soil screening for DDT by immunoassay	8081	Greater or less than 0.2, 1.0, and 10 mg/kg
4050—TNT explosives in soil by immunoassay	8330	0.5 to 5.0 mg/kg
4051—RDX in soil by immunoassay	8330	0.5 to 6 mg/kg
6200—Field portable x-ray fluorescence spectrometry for the determination of elemental concentrations in soil and sediment	6010, 6020, 7000	10 to 200 mg/kg
8510—Colorimetric screening method for RDX in soil	8330	1 to 30 mg/kg
8515—Colorimetric screening method for TNT in soil	8330	1 to 30 mg/kg

2,4-D denotes 2,4-dichlorophenoxyacetic acid.
DDT denotes 1,1,1-trichloro-2,2-bis(4-chlorophenyl) ethane.
TNT denotes trinitrotoluene.
RDX denotes Royal Demolition Explosive.

3.9.3.1 Immunoassay screening kits

Immunoassay test kits are manufactured in several different formats for rapid analyses of gasoline, diesel fuel, organochlorine pesticides, herbicides, and explosives with EPA-approved methods listed in Table 3.9. Depending on the format, the kits may produce quantitative or semiquantitative data. With appropriate training and experience, the kits are relatively easy to use. The manufacturer provides step-by-step instructions for the kit operation; these procedures must not be modified under any circumstances. Samples are analyzed in batches; the analysis of one batch may take 45–60 minutes (min). The kits contain the necessary prepackaged reagents, glassware, tools, and a portable spectrophotometer. Immunoassay kits have a limited shelf life and must be refrigerated per manufacturer's instructions. They cannot be exposed to direct sunlight and must not be used at temperatures exceeding 81° Fahrenheit (F) or 27°C.

Unlike laboratory analysis, the majority of immunoassay kits do not produce results that are expressed as single concentrations. Results may be expressed semi-quantitatively as the ranges of concentrations within which the contaminant concentration falls (for example, greater than 0.5 mg/kg and below 2 mg/kg), or as comparisons to a standard value, such as less than 1 mg/kg or greater than 1 mg/kg. Immunoassay kits produce quantitative data only when configured with a series of reference standards used to construct a calibration curve.

Immunoassay screening kits developed for a specific compound have cross-reactivity with compounds of the same chemical class and may produce false positive results. False negative results may also occur. They are particularly frequent in petroleum fuel screening because the natural attenuation of petroleum fuels in the environment causes the depletion of fuel constituents that are critical in the immunoassay testing. When using field screening kits, at the beginning of the project we must establish a correlation between the screening results and definitive laboratory analyses with EPA methods listed in Table 3.9, and continue to confirm the field data with laboratory data for at least 10 percent of all samples.

3.9.3.2 XRF screening

Soil screening for metal contaminants with an x-ray fluorescence spectrophotometer is a technique used in EPA Method 6200. The method is applicable for 26 analytes listed with their detection limits in an interference-free matrix in Table 3.10. Light elements, such as lithium, sodium, aluminum, silicon, magnesium, beryllium, phosphorus, cannot be detected with XRF.

When samples are irradiated with x-rays, the source x-rays are absorbed by the atoms of metals in the sample. When an atom absorbs the source x-rays, the radiation dislodges the electrons from the innermost shells of the atoms, creating vacancies. The vacancies are filled with electrons cascading from the outer electron shells. Because electrons in the outer shells have a higher energy state compared to electrons in the inner shells, they give off energy as they fill the positions in the inner shells. This rearrangement of electrons results in emission of x-rays that are characteristic of a given atom and produces an x-ray fluorescence spectrum.

XRF analysis may be used in two modes: *in situ* and *ex situ*. To take an *in situ* measurement, the window of a hand-held field portable XRF instrument is pressed

Table 3.10   Detection limits for XRF analysis in interference-free matrices[1]

Element	Detection limit (mg/kg)	Element	Detection limit (mg/kg)
Antimony	40	Nickel	50
Arsenic	40	Potassium	200
Barium	20	Rubidium	10
Cadmium	100	Selenium	40
Calcium	70	Silver	70
Chromium	150	Strontium	10
Cobalt	60	Thallium	20
Copper	50	Thorium	10
Iron	60	Tin	60
Lead	20	Titanium	50
Manganese	70	Vanadium	50
Mercury	30	Zinc	50
Molybdenum	10	Zirconium	10

[1] EPA, 1998f.

against the surface of the soil to be screened. In *ex situ* XRF analysis, a small plastic cup is filled with a dried, ground, and sieved soil sample and placed inside the instrument.

The XRF analysis accuracy depends on physical properties of the soil, such as particle size, moisture content, homogeneity, and surface conditions. Drying, grinding, and sieving soil prior to analysis reduce the effects of these physical interferences.

Chemical interferences may be produced by overlapping spectra of different elements or as the result of x-ray absorption or enhancement. Either effect is common in soil contaminated with heavy metals. Chemical interferences may be substantially reduced through a mathematical correction, but they cannot be completely eliminated. Other factors that affect the accuracy of XRF analysis are the instrument settings and the operator technique, especially in *in situ* measurements. A correlation of XRF results with laboratory analysis by other analytic techniques should be always established in the early stages of the project implementation and confirmed, if changes in the nature of soil samples have been observed.

Physical and chemical interferences elevate the XRF analysis detection limits; for certain soil types, the actual detection limits may exceed these for an interference-free matrix shown in Table 3.10. The XFR detection limits are always higher than the detection limits for other elemental analysis techniques. The high detection limits may not be appropriate for some types of DQOs and are the greatest limitation of XRF screening.

Practical Tips: Field screening kits

1. Select field screening kits carefully; discuss their applicability and limitations with the manufacturer's technical representative.

2. Contact the manufacturer's sales representative for a specific quote to obtain a volume discount price.

3. The spectrophotometer kit used in immunoassay testing is the most expensive part of the test kit hardware. It may be rented from the manufacturer.

4. The EPA approval does not guarantee that a screening method is going to work on a specific matrix.

5. Whenever possible, confirm screening results with definitive data for at least 10 percent of all samples.

6. Field screening should be conducted in an area protected from sunlight, wind, and rain.

7. Make sure that only properly trained personnel conduct testing with field screening kits.

8. Keep good records of field screening measurements.

3.10 FIELD RECORDS

Field records document field measurements and field conditions at the time of sampling. These are permanent project records that are stored with the rest of the project files in a central location after the project is completed. A chemist, a geologist, or an engineer often reviews the field records to order to interpret the collected data in context of field conditions during sampling. Accurate and relevant field records allow efficient resolution of issues related to field and laboratory data interpretation and quality, whereas sketchy and inaccurate records raise more questions than they answer.

At a minimum, field records include the COC Form and field logbooks. Matrix-specific sampling forms, similar to one shown in Appendix 16, may often supplement logbooks. Another form of field records is a *photolog* that may be created for certain projects.

The following common rules apply to all field records:

- All entries are made in indelible, non-running black ink (no pencil use)
- Empty pages marked with diagonal line and initialed
- Each page and end-of-day page are signed and dated
- Errors marked out with single diagonal line with initials

Chain-of-Custody Form

The COC Form, which is started in the field and completed at the laboratory, serves as a legal record of sampling and as a request for analysis. It contains only the basic field information that is necessary for the laboratory to correctly identify and analyse the samples. An incomplete or incorrect COC Form usually causes delays at the laboratory, which in turn may jeopardize the sample holding time and the turnaround time of analysis.

Field Logbooks and Sampling Forms

Field data and observations are recorded in field logbooks and sampling forms. Field logbooks are bound notebooks with numbered pages. Each project should have its own dedicated logbook; large projects with several sampling crews may have several field logbooks. The following information is recorded in field logbooks:

- Project name and location
- Name, title, and affiliation of all personnel on-site, including the note taker
- Weather information for the day of sampling (approximate temperature; wind speed and direction; humidity; precipitation)
- Description of work performed (excavation, borehole placement, groundwater monitoring, etc.)
- A map showing sampling points (groundwater well field; a grid from which samples were collected; a drawing of an excavation with measurements; etc.)
- Date and time of sampling for each sample
- Sample matrix (soil, water, wipe, etc.)
- Sample number
- Sampling point identification (soil boring, grab sample location, monitoring well ID, etc.)
- Depth at which the sample was collected
- Sample type (composite or grab)
- Composite description (the number of grab samples)
- Analysis requested for each sample
- Field measurements performed (calibration records, field parameter measurements, field screening data)
- Deviations from the SAP
- Problems encountered and corrective actions taken (sampling problems, alternative sampling methods used)
- QC and QA samples and the corresponding primary samples
- Instructions received in the field from the client or the oversight agency
- Observations (presence of sheen on the sampled water; unusual sampling point or sample appearance; the formation of effervescence in preserved water samples; change in weather conditions; etc.)

Waterproof and regular paper field logbooks are available from several sampling supply distributors; bound school notebooks with numbered pages will work equally well. For long-term projects, we may create custom logbooks with project-specific information and sampling forms preprinted and bound into paginated notebooks. Hand-held data logging computers may be also used, with the data uploaded into a database for permanent storage.

The use of sampling forms does not eliminate the need for a field logbook. Sampling forms are typically designed to document a specific sampling activity at one sampling point (groundwater sampling, well logging) and they do not have the room for recording the field observations and other important information, such as field instrument calibrations, client's instruction, field variances, etc. When sampling forms are used, limited sampling information is recorded in the form, whereas the rest of the information is documented in the field logbook.

Photolog
Photologs provide a visual record of project activities and document field conditions that can be best described by a picture. Not every project requires a photolog. For example, groundwater sampling does not need it (unless a damaged wellhead is to be documented), but UST and pipeline removals, soil excavations, and building demolition projects do. Leaking USTs with gaping holes, stained soil or wall surfaces are best documented visually. Photologs provide the information that is supplemental to the field logbook record and enable the viewers to form their own opinion on field conditions described in the field logbook. Photologs are made with digital cameras, and the captioned images are usually included in the project report as an appendix.

4

Understanding the analytical laboratory

The second half of the implementation phase of the environmental chemical data collection process consists of Task 4—Laboratory Analysis and Task 5—Laboratory QA/QC. Arising from a foundation of project planning and following field sampling implementation as shown in Figure 4.1, these tasks transform the collected samples into chemical data.

The analytical laboratory with its dozens of seemingly similar methods and arcane QA/QC procedures is a mysterious world that most professionals who work in consulting industry rarely visit or fully comprehend. The perplexing terminology, the subtleties of analytical QA/QC protocols, and the minutia of laboratory methods usually intimidate those who come in contact with the analytical laboratory only occasionally. However, the knowledge of the processes that take place during laboratory analysis would enhance any practicing professional's expertise. This chapter is not intended as a course in analytical chemistry, environmental analysis or laboratory management. Its purpose is to broadly describe laboratory operations, identify the critical aspects of chemical analysis, and emphasize their importance for data quality.

Modern laboratories face enormously complex qualitative and quantitative analytical problems in their daily work. Using various analytical techniques, they routinely identify and quantify hundreds of organic and inorganic analytes in an assortment of environmental matrices while adhering to the requirements of specific QA/QC protocols. All too often, the *perceived* validity of obtained data is based on the observance of these specific requirements, whereas the *actual* validity of the data depends only on solid qualitative and quantitative analysis and on the adherence to standard analytical laboratory QA/QC protocols.

To fully appreciate the intricacies of analytical laboratory work, one must spend many years performing various laboratory functions and tasks. In fact, a full service analytical laboratory is such a complex organization that even some of the laboratory personnel may not fully understand the relationships between different laboratory functions or production units.

4.1 REQUIREMENTS FOR LABORATORY OPERATION

The current standard for environmental laboratory operation is a combination of several existing standards. General requirements for testing laboratory operations are

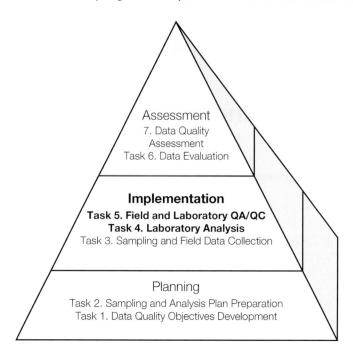

Figure 4.1 Laboratory implementation phase of environmental chemical data collection.

described in the International Standardization Organization (ISO) Standard 17025 (ISO, 1999), formally known as ISO Guide 25 (ISO, 1990), and in the Test Methods for Evaluating Solid Waste, SW-846 (EPA, 1996a). The ISO Standard 17025 requirements apply to laboratories that perform any kind of calibration and testing, whereas the SW-846 requirements are specific to environmental laboratory operation. To enforce a minimum standard, most of the states within the USA have their own environmental laboratory accreditation programs. The rigorousness of these programs varies from state to state, with some being extremely demanding and others very lax. This multiplicity creates a difficult business climate for environmental laboratories, as working in many states entails maintaining multiple accreditations at significant effort and expense.

Recognizing the need for uniform nation-wide standards for environmental laboratory operations, the EPA together with other federal agencies, the states, and the private sector representatives created the National Environmental Laboratory Accreditation Conference (NELAC). This organization has developed the standards for laboratory operation that have been published in NELAC's Program Policy and Structure. This document undergoes frequent revisions; the latest revision can be found on NELAC's web page www.epa.gov/ttnnela1. The NELAC standards for laboratory operation, which are based on the ISO Guide 25 requirements, are the most comprehensive standards that exist today.

The NELAC standards serve as criteria for the National Environmental Laboratory Accreditation Program (NELAP), which so far has operated on the basis of voluntary laboratory and state participation. The intent of the NELAP is to oversee the implementation of the NELAC standard and to accredit laboratories under it. As the states subscribe to the NELAC standard and request that laboratories be accredited through the NELAP, the need for state accreditation will disappear, and all laboratories in the USA will eventually operate to a single comprehensive standard.

Because the NELAC standards are general, they require interpretation or clarification in certain areas of application. That is why the US DOD undertook a task of specifying some of these broad areas in definite terms in a manual that unifies laboratory requirements of the Navy, Air Force, and United States Army Corps of Engineers (DOD, 2000). This manual, titled *Department of Defense Quality Systems Manual for Environmental Laboratories*, is offered to the DOD representatives and contractors who design, implement, and oversee contracts with environmental laboratories. Currently, the three DOD branches implement their own laboratory accreditation programs with some reciprocity between the three; a task force has been created to combine these programs into a single DOD accreditation process.

Another standard that influenced environmental laboratory operation is the Federal Insecticide, Fungicide, and Rodenticide Act (FIFRA) Good Laboratory Practice (GLP) Standards (EPA, 1989b; EPA, 1998e). These standards were formulated as the general requirements for the laboratories that specialize in pesticide testing. The FIFRA became law in 1972 and enabled the federal government to control pesticide distribution, sale, and use. The GLP Standards are mandatory for laboratories that support pesticide registration and application in the USA. Although the FIFRA does not control laboratories that perform analysis of environmental matrices, the provisions of the GLP Standards had a profound effect on the operation of environmental laboratories. The GLP Standards implementation became a requirement for the laboratories that provide analysis under negotiated testing agreements for the TSCA studies on health and environmental effects of toxic chemicals and their fate (EPA, 1998c). A vast majority of environmental laboratories adopted many provisions of the GLP Standards to ensure scientific reliability of the produced data and to create a documentation system that allows a complete reconstruction of results from laboratory records.

Based on the provisions of the GLP Standards and the policies on information management and computer security, the EPA developed a set of procedures for reliable and secure use of laboratory information management systems (LIMS). These procedures are summarized in the Good Automated Laboratory Practice (GALP) guidance (EPA, 1995b). Although the GALP is not a mandatory standard, the laboratories that are concerned with the integrity of their data recognize its value and implement the GALP entirely or in part.

As one can see, the current laboratory operation standard is a loose compilation of several existing standards. As a result, each laboratory operates to a set of its own rules that have emerged over the years from the requirements of different accrediting entities. This variability makes the understanding of internal laboratory process difficult. More complications are introduced by a great variety of analytical methods and the possibilities for their interpretation. Specific guidance for laboratory analyses has been promulgated under various EPA laws, such as the CERCLA, RCRA, SDWA, and CWA. The laboratories interpret the somewhat ambiguous language of these methods in a variety of ways, thus creating further inconsistency. The implementation of PBMS in the analytical laboratory introduces another complication for data users in understanding the laboratory process.

There is, however, a set of basic rules that every laboratory must follow in order to produce technically and legally defensible data, and these rules are discussed in the following chapters.

4.2 LABORATORY ORGANIZATION

Different laboratories have different internal organizations, but all laboratories have two things in common: the general process by which environmental samples are transformed into data, and a laboratory quality system.

The general process of sample transformation into data is the same at every environmental laboratory. The differences are primarily in the manner various tasks are performed by laboratory personnel. To assure the quality of produced data, every laboratory must develop, implement, and maintain a quality system that is documented in the Laboratory QA Manual. The implementation of specific tasks related to sample management, analysis, and quality system, which may be different at different laboratories, is addressed through a set of laboratory's own SOPs. A full service laboratory has dozens of SOPs, describing every laboratory procedure and task from sample receiving to invoicing. The SOPs are updated as necessary and undergo internal review and approval.

To better understand the structure and the inner workings of an environmental laboratory, we need to familiarize ourselves with laboratory functional groups and their responsibilities. Figure 4.2 shows an example of a typical full service environmental laboratory organization chart. A full service laboratory has the capabilities to perform analysis for common environmental contaminants, such as VOCs and SVOCs (including petroleum fuels and their constituents, pesticides, herbicides, and PCBs), trace elements (metals), and general chemistry parameters. Analysis of dioxins/furans, explosives, radiochemistry parameters, and analysis of contaminants in air are not considered routine, and are performed at specialized laboratories.

As shown in Figure 4.2, a typical laboratory organization consists of several functional groups or sections that are summarily called Laboratory Operations or just Operations. These are the groups performing different types of analyses, such as Organic and Inorganic Analysis Sections. The support groups in Figure 4.2 are named after their respective functions: Sample Receiving, Data Management, Customer Service, and Administrative Support. Separate from Operations is the Quality Assurance Section that provides independent QA oversight to Operations. The Laboratory QA Manager reports directly to the person in charge of the laboratory and is not involved in Operations. Small laboratories with limited resources may not have all of the functional groups shown in Figure 4.2, but even the smallest laboratory must have independent QA oversight.

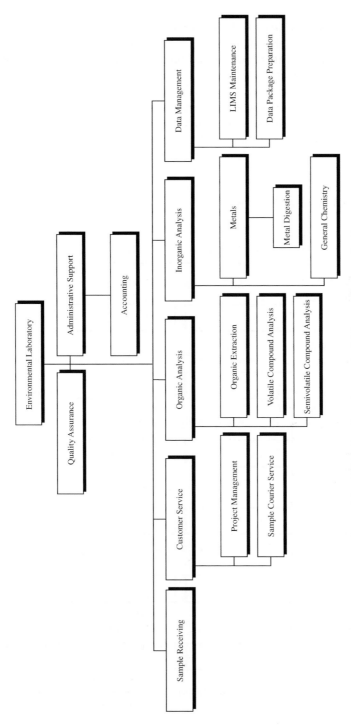

Figure 4.2 Example of environmental laboratory organization.

4.3   THE SEVEN STEPS OF LABORATORY PROCESS

Similar to the seven steps of the DQO process, the general laboratory process of sample transformation into data can be broken down into seven typical steps shown in a diagram in Figure 4.3.

Each step signifies a milestone in the sample's travel through the laboratory. Unlike other seven-step diagrams with irreversible flow that we have seen in previous chapters, a unique feature of laboratory process is that the relationships between individual steps are for the most part reversible. This means that there is a viable chance to correct an error made at the previous step. This reversibility is the greatest advantage of the laboratory process. As laboratories handle thousands of samples and analyze them by dozens of various methods, the possibility for making an error at any step is real, but so is the opportunity to correct the error. Obviously, some errors

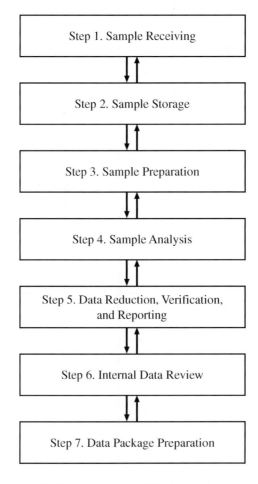

Figure 4.3   The seven steps of the laboratory process.

cannot be remedied, particularly in cases when the sample has been used up in analysis, but overall, if timely discovered, most of laboratory errors are correctable.

There are several critical areas in the laboratory process, which have a potential to completely invalidate a sample in a technical sense and from a legal perspective. Because the laboratory process is poorly understood by most of the outsiders, we will review every step of it in order to comprehend the relationships between different steps and to recognize the effects each one has on data quality.

4.3.1   Step 1—Sample receiving

Samples are received at the laboratory by a group of people designated as Sample Custodians who work within the Sample Receiving Section. This section is typically responsible for entering the COC Form information into the LIMS; sample storage; the preparation of sample containers and coolers on client's request; and the disposal of expired samples. These procedures must be documented in laboratory SOPs that are kept readily available at the sample receiving area.

Sample Custodians open the coolers, remove the COC Forms and verify the COC Form against the cooler contents. At many laboratories, Sample Custodians will not sign and date the COC Form and log-in a sample into the LIMS until all identified discrepancies have been resolved with the client. The resolution of outstanding issues may delay the start of analysis and erode the sample holding time. The following mistakes made by field sampling personnel are detected at this first step of the laboratory process:

- Missing samples
- Incomplete, unclear COC Forms
- Broken or leaking sample containers due to poor packaging
- Unidentified samples because the labels had peeled off under the action of moisture
- Unreadable labels that were made with runny ink and not protected with clear tape
- Incorrect sample volumes, containers or preservation

Sample Custodians determine from the COC Form whether a temperature blank has been enclosed with the samples. If a temperature blank has been enclosed, Sample Custodians remove the ice, the packing material, and the samples from the cooler and line them up on a receiving table or in a fume hood, if samples emanate odor. They measure the cooler temperature by inserting a thermometer into the temperature blank. If the blank is not present, they measure the temperature inside the cooler by placing a thermometer or an infrared temperature probe between sample containers. The temperature of the cooler upon arrival to the laboratory should be 2–6°C whether it has been measured inside the cooler or in the temperature blank. To document sample conditions upon arrival at the laboratory, laboratories record the cooler temperature on the COC Form or use a separate cooler receipt form, similar to one shown in Appendix 17. Samples are then placed in storage refrigerators or walk-in coolers kept at 2–6°C.

After all of the samples have been matched to the COC Form, Custodians Sample log them into the LIMS. Each group of samples that came under one COC Form, single or multi-page alike, is assigned a unique identifying number, called a *Work*

Order or a *Laboratory Project Number*. The LIMS automatically assigns a sequential laboratory sample ID number to each sample and prints laboratory labels for Sample Custodians to place on the sample containers.

Based on the analytical requirements of the COC Form, Sample Custodians assign the samples with internal laboratory codes for the preparation and analysis and enter into the LIMS the turnaround time of analysis requested by the clients. At some laboratories, the LIMS may create internal work orders for use by other laboratory sections and print internal custody forms for tracking samples inside the laboratory.

Many laboratories have a Customer Service Section, staffed with Project Managers who serve as points of contact with the clients. As part of customer services, many laboratories offer a free sample courier service to the projects that are located within a reasonable driving distance. The Customer Service Section works in close contact with the Sample Receiving Section. Project Managers usually review and verify the correctness of their clients' laboratory work orders generated by Sample Custodians and resolve issues that are associated with sample shipments. They are responsible for managing all the needs of their clients, including scheduling of upcoming analytical projects; arranging the delivery of sample containers and coolers to project sites; sample pick-up courier services; change orders; special analytical requirements and the resolution of technical issues.

Prior to the start of sampling we must make sure that the laboratory is prepared to meet the project-specific requirements contained in the SAP. We should always send the SAP to the Project Manager, who in turn will ensure that it is available to laboratory personnel. A meeting or a conference call with the Project Manager and other key laboratory staff prior to start of field work is always beneficial for establishing a common purpose and for building a relationship with the laboratory.

Sample receiving is not an error-free process; it requires attention to detail and a good knowledge of analytical methods. Common sample receiving errors are discussed in Example 4.1.

Example 4.1: Sample management errors during sample receiving

Sample management errors made in the course of sample receiving affect mostly the collected sample representativeness. Laboratories successfully prevent most of errors in sample receipt and storage by implementing QA measures, such as the sample storage at 2–6°C; the segregation of samples to prevent cross-contamination; and the adherence to approved SOPs for sample management. Errors in sample labelling and internal work orders may be detected and corrected during internal review of the COC Forms and internal laboratory documentation, a rule implemented at many laboratories. The only irrevocable errors that may take place at this step are a loss of samples due to negligent handling and an undetected error in assigning an improper analysis code in the LIMS followed by a wrong analysis.

4.3.2 Step 2—Sample storage

Samples are stored in laboratory refrigerators or walk-in coolers that are kept at 2–6°C. Water samples collected for metal analysis are stored at room temperature at some laboratories and in cold storage at others, reflecting the conflicting CLP and

that specialize in these two branches of analytical chemistry. The Organic Analysis Section may be divided into VOC and SVOC Analysis Groups and the Organic Extraction Group, which prepares samples for SVOC analyses. The Inorganic Analysis Section may be divided into Metal Analysis Group, which includes the Metal Digestion Group, and the General Chemistry Group.

Every analyst within an analytical group must be trained in the procedures they conduct, and their proficiency must be documented according to the Laboratory QA Manual. The laboratory must have current SOPs for all performed analytical methods.

The state-of-the-art laboratories are equipped with the latest models of analytical instruments and computer systems, while others may have older, less sophisticated equipment or a mix of modern and outdated instruments. The goal of production laboratories is to analyze samples in the fastest possible manner. To be competitive, laboratories must have fully automated analytical systems allowing unattended sequential analysis of samples and computerized output of analytical results. Data acquisition computers, programmed with specialty software, control analytical instruments, collect the raw data, and convert them into analytical results. These computers are typically interfaced with the LIMS, which networks different laboratory sections into a single computer system and transforms analytical results into laboratory reports.

After analysts have determined from the LIMS (or by other means of communications) that samples have been prepared and ready for analysis, they retrieve extracts or digestates from storage and assemble an analytical batch. *An analytical batch is a group of samples, extracts or digestates that are analyzed sequentially using the same instrument calibration curve and that have common analytical QC checks*. The number of samples in an analytical batch is not limited.

The analysts calibrate the instruments by preparing calibration standards at several known concentrations and analyzing them according to the method and laboratory SOP requirements. They cannot begin analyzing samples until the initial calibration meets the acceptance criteria specified for each target analyte. Organic analyses typically have 5-point calibration curves, whereas inorganic parameters may have 3- to 5-point calibration curves.

After initial calibration has been completed, it must be verified with a *second source standard*, which has been prepared from material coming from a different source (different manufacturer or lot number). This QC measure, called the *second source confirmation*, allows detecting errors in the initial calibration standard solution preparation and prevents a systematic error in compound quantitation. Only after this *initial calibration verification* (ICV) standard has also met its own acceptance criteria may the analyst start analyzing samples.

The calibration of instruments used for organic compound analyses is a particularly lengthy procedure due to the large number of target analytes that are analyzed simultaneously. For example, laboratories analyze a minimum of 55 chemicals in EPA Method 8260 and a minimum of 75 chemicals in EPA Method 8270. (These numbers may be as high as 100 for either method). Even with the help of modern computers, a calibration procedure of this enormous scope may easily take 6–8 hours or even longer in some circumstances. Fortunately for the laboratories, once these initial calibrations have been established, they may stay valid for long periods of time, sometimes as long as 3–6 months. To use them for quantitation, the laboratories must daily verify their validity by analyzing a calibration verification standard. This calibration check, called **continuing calibration verification** (CCV), must meet the method and the laboratory SOP acceptance criteria for the analysis to proceed. The CCV is analyzed daily at a method-specified frequency. (Chapter 4.5.2 details calibration requirements in environmental analysis.)

After the CCV acceptance criteria have been met, the analyst conducts a series of tasks in preparation for sample analysis, such as the following:

1. **Organize Analytical Batches**—These include analytical QC checks (instrument blanks and other QC checks as required by the method), CCV standards, prepared field samples, and laboratory QC samples.
2. **Add Internal Standards**—As part of internal standard calibration procedure, the analyst adds internal standards to the calibration standards and sample extracts prior to analysis.
3. **Prepare the Analytical Sequence**—Using the computer that controls the analytical instrument, the analyst creates an analytical sequence, which is a list of all samples in the analytical batch. The entered information for each sample may include laboratory sample ID, the volume or weight used for preparation, the final extract volume, and a dilution factor, if the extract was diluted.
4. **Load the Autosampler**—The analyst then places the samples into the autosampler and programs the instrument to start analysis.

To maximize instrument use, laboratories prefer to analyze large batches of samples. An average batch is about 20 samples large, however batches of up to 50 samples are not uncommon. Depending on the laboratory's implementation of a standard method, one organic analysis may take 30–60 minutes. That is why large analytical sequences for organic analyses are usually analyzed overnight. Analysis for one trace element is a matter of approximately 3 minutes, allowing fast throughput of samples and analysis during laboratory working hours. Inorganic analyses are for the most part conducted manually with the data recorded by hand in laboratory notebooks, and they may be lengthy and extremely laborious.

After a sample sequence has been analyzed, the analyst evaluates a computer printout with the raw analytical data and the computer-calculated analytical results. The analyst verifies that the QC acceptance criteria for daily calibration verifications, the surrogate and internal standards, and laboratory QC samples have been met. This review enables the analyst to determine whether reanalysis of samples will be required. Sample dilutions may be also needed if analyte concentrations in samples exceed the instrument calibration range.

The analyst's interpretation skills and solid judgment are critical for analyses where pattern recognition is required (e.g. petroleum fuels or PCBs) and in discerning the presence of false positive or false negative results or the effects of matrix interferences. After having identified samples for reanalysis, the analyst begins organizing another analytical batch, and the new analysis cycle begins.

The raw data and the hardcopy analytical results for original analysis and all reanalyses become part of the sample file. They are maintained as a permanent analytical record and may become part of a data package deliverable to the client. The sample and laboratory QC sample results are transferred into the LIMS for the generation of final analytical reports.

Common errors that may be made during analysis are measurement and procedure errors; the most typical ones are featured in Example 4.5.

Example 4.5: Errors in analysis

Measurement errors
The laboratory detects measurement errors by comparing the accuracy and precision of sample and laboratory QC check sample analysis to acceptance criteria. If results do not meet the criteria, the laboratory identifies the reason and takes corrective action, such as the following:

- Reanalysis of samples or extracts
- Re-extraction (or redigestion) and reanalysis
- Creation of a new multipoint calibration followed by sample reanalysis

Procedure errors
The following errors of procedure may be identified in the course of raw data review:

- Method blank contamination
- Analytical instrument contamination from analysis of samples with high contaminant concentrations (memory effects or carryover)
- Sample cross-contamination in cold storage and in preparation

Another procedure error is the lack of documentation during analysis that hinders the reconstruction of results during data evaluation in the assessment phase of the data collection process. Substituting analytical methods without client's consent is another procedure error that may have far reaching consequences for data use.

Much more difficult to detect are data interpretation and judgment errors, such as the unrecognized false positive and false negative results and the incorrect interpretation of mass spectra and chromatographic patterns. The detection and correction of these errors is made possible through internal review by experienced analysts.

The worst kind of error that can take place during analysis is the use of analytical methods that have not been validated by the laboratory to establish their performance. Inadequately validated methods, published and non-standard alike, are not considered to be sufficiently reliable to produce data of known quality. The laboratory must validate every analytical method, whether it is an approved method or a non-standard PBMS developed by the laboratory.

4.3.5   Step 5—Data reduction, verification, and reporting

Data reduction is the mathematical and statistical calculations used to convert raw data to reported data. *Data verification* is the process of confirming that the calculations used in data reduction are correct and that obtained data meet the applicable analytical requirements.

In a modern laboratory, automated computer software for data acquisition and processing performs most of data reduction. Raw data for organic compound and trace element analyses comprise standardized calibration and quantitation reports from various instruments, mass spectra, and chromatograms. Laboratory data reduction for these instrumental analytical methods is computerized. Contrary to instrumental analyses, most general chemistry analyses and sample preparation methods are not sufficiently automated, and their data are recorded and reduced manually in laboratory notebooks and bench sheets. The SOP for every analytical method performed by the laboratory should contain a section that details calculations used in the method's data reduction.

In order to produce defensible data of known quality, in addition to method-specific SOPs, the laboratory should develop a Software QA Manual, which describes activities related to data generation, reduction, and transfer with modern tools of data management, and the policies and procedures for modification and use of computer software.

Data reduction, verification, and reporting are always a combined effort of the personnel working in different laboratory sections, their supervisors, and the Data Management Section. Sample preparation technicians and analysts performing analysis are responsible for the correctness of their computerized and manual calculations and for proper reporting of the obtained results. Supervisors verify the correctness of these calculations during data review and approval process. The Data Management Section staff is responsible for the correct conversion of the data entered by hand or transmitted from the data acquisition computers into the LIMS and the production of finished laboratory reports prepared to client's specifications.

The use of computers enables laboratories to greatly improve their production capabilities, but it also creates new tasks and responsibilities. For example, laboratories must verify computer programs used for data reduction in order to confirm the correctness of the calculations. To create an analytical method using commercially available or custom analytical software, laboratory chemists must exercise professional judgment in the selection of appropriate method parameters often accompanied by data entry by hand, which opens the door for errors to creep in. Today, electronic spreadsheets replaced manual calculations, and their correctness must be also verified.

The laboratory should verify and document the proper functioning of the software immediately after any new data acquisition or management systems have been installed. The baseline verification consists of manual calculations to confirm the correctness of all computer calculations. Ongoing verification takes place during laboratory data review process whenever a reviewer replicates one of the results generated by the computer or a manual calculation from a bench sheet. All information used in the calculations (raw data, calibration data, laboratory QC checks, and blank results) is kept on file for the reconstruction of the final result at a later date, should it become necessary. Bench sheets that document sample preparation are also kept on file for the same purpose.

Example 4.7: **Rules for significant figure determination**

Result of Calculation	Number of Significant Figures	Consideration of Zero
0.0011	2	The three zeros serve to position the decimal point and do not constitute significant figures.
11×10^{-2}	2	The 10^{-2} serves to position the decimal point and does not constitute significant figures.
1.1	2	No zeros to consider.
1.100	4	The two zeros do not define the position of the decimal point and are significant figures.
11	2	No zeros to consider.
11.00	4	The two zeros do not define the position of the decimal point and are significant figures.
110	3	The zero does not define the position of the decimal point and is a significant figure.
110.1	4	The zero does not define the position of the decimal point and is a significant figure.

In the data reduction and reporting steps, as computers do the actual calculations and the significant figure rounding off, the laboratory staff should program them according to the rules for significant figure determination. Many laboratories carry one extra significant figure throughout all calculations, and round off the final result to the number of significant figures appropriate for the accuracy of the analytical method.

Common conventions for reporting of environmental sample analytical results are as follows:

- Most laboratory results are reported in two significant figures.
- Laboratory QC sample results are often reported with three significant figures. This does not, however, mean they have a higher accuracy than environmental samples.

4.3.5.4 Rules for rounding off
When calculations by a computer or by hand produce more figures than needed, they are rounded off to the proper number of significant figures. Laboratories commonly use the following rules for rounding off:

1. If the next digit beyond the rounding point is less than 5, leave the previous digit unchanged (e.g. 11.3 becomes 11).
2. If the next digit beyond the rounding point is greater than five, increase the previous digit by one (e.g. 11.6 becomes 12).
3. If the next digit beyond the rounding point is a 5 followed by other digits, then treat the case as for greater than five as in Rule 2 (e.g. 11.51 becomes 12).

4. If the next digit beyond the rounding point is equal to 5 with no digits other than zeros following the 5, round the previous digit to the nearest even number (e.g. 11.50 and 12.50 both become 12).

Example 4.8 illustrates the rules for rounding off and the significant figure determination.

Example 4.8: Rounding off

- By multiplying 1.2345 (five significant figures) by 0.5 (one significant figure) we will obtain a result of 0.61725 and round it off to 0.6 (one significant figure).
- By multiplying 1.2345 (five significant figures) by 0.50 (two significant figures), we will obtain a result of 0.61725 and round it off to 0.62 (two significant figures).

4.3.5.5 Moisture correction

Analytical results for solid samples may be reported either on a dry weight basis or on a wet weight basis ('as is'). Soil and sludge results reported on a dry weight basis determine contaminant concentrations in the sample's solids.

A sample aliquot is dried at 105°C, and a change in weight due to the loss of moisture is recorded. Percent moisture and percent solid are calculated as follows:

$$\text{Percent moisture} = \frac{W_W - W_D}{W_W} \times 100$$

where W_W is the weight of a sample before the drying and W_D is the weight of a dried sample.

$$\text{Percent solid } (P_S) = 100 - \text{percent moisture}$$

The calculation to determine the final dry weight contaminant concentration in a soil sample is as follows:

$$C_D = \frac{C_W}{P_S} \times 100$$

where C_D is the concentration corrected for dry weight and C_W is the uncorrected (wet weight) concentration.

While the dry weight correction may be a necessary procedure for monitoring of some industrial sludges with variable moisture contents, the reason for dry weight reporting of soil samples is not always clear. This practice could have originated from risk assessment, where calculations have been traditionally done using moisture-corrected results. (This practice is, however, changing, and today some risk assessment calculations are done based on contaminant concentrations not corrected for moisture content). Dry weight correction is a required procedure under some analytical protocols (EPA SW-846 and EPA CLP SOW) and is a requirement of the DOD contract work.

Another reason for dry weight correction may be a need for common grounds when comparing analytical results to action levels. Action levels (PRGs, RBCA criteria, or petroleum fuel cleanup goals) must be based on dry weight concentrations in order to be compared to contaminant concentrations in environmental samples that have varying moisture contents. Analytical results reported on a dry weight basis are greater than the same results uncorrected for moisture content, providing more conservative estimates of true contaminant concentrations.

4.3.5.6 Reporting of undetected compounds

A vast majority of all reported data are results for chemical pollutants that have not been found in the analyzed samples. There are three different reporting conventions for compounds that have been analyzed for but have not been detected. These conventions are summarized in Table 4.2.

Some laboratories report undetected analytes using the acronym 'ND,' i.e. 'Not Detected,' placed next to the analyte name in the report. If the PQL for the analyte is also shown in the report, we will know that this analyte was not detected at a concentration that is equal to or greater than the analyte's PQL. If a sample has been diluted for analysis, the reporting limits for the analytes in this sample become elevated. The dilution factor (DF) must also be included into the report to enable us to calculate the reporting limit for every analyte by multiplying the PQL by the dilution factor. This kind of reporting does not make the reading and understanding of analytical reports easy, especially if vital information, such as the PQLs or the DFs, is missing or incomplete. (The definitions of the PQLs and RLs are given in Chapter 4.5.1).

An example of such reporting and the associated calculations (shown in the first row of Table 4.2) demonstrates that this reporting convention is not informative enough for practical use because the results for either diluted or undiluted samples are reported similarly as 'ND.' Although laboratories usually indicate the corresponding reporting limits on the report of analysis, this reporting convention is flawed because data reported as 'ND' cannot be used for comparisons and calculations, unless they are transformed into a numeric format. (As a side note, this reporting convention gave birth to the colloquialism of 'Non-detect,' which is often carelessly put in technical writing in a verb form.)

The second reporting convention for undetected analytes uses a '<' (less than) symbol preceding the PQL or RL value as shown in the second row of Table 4.2. This

Table 4.2 Reporting conventions for undetected analytes

Analyte	PQL (mg/kg)	Result for undiluted sample (mg/kg)	DF	RL[1] for diluted sample (mg/kg)	Result for diluted sample (mg/kg)
Benzene	0.005	ND	100	0.5	ND
Toluene	0.005	<0.005	100	0.5	<0.5
Ethylbenzene	0.005	0.005U	100	0.5	0.5U

[1] RL = PQL × DF

type of reporting is more informative than the first one, as it instantly provides the necessary information on the numerical values of the PQL or RL.

The third reporting convention is the one of EPA CLP SOW, which requires that laboratories report data for undetected analytes by indicating the analyte-specific PQL or RL accompanied by the qualifier 'U' (undetected), as shown in the third row of Table 4.2. This convention has been integrated into the DOD analytical protocols and became a standard reporting format at many laboratories.

The advantages of the two latter kinds of reporting conventions are the immediate availability of the PQL or RL information and their instant comparability with numeric action levels.

4.3.6   Step 6—Internal data review

Detailed review of reported data reduces laboratory risk of producing invalid data. Important features of internal data review are the spot checks of calculations; the verification of the acceptability of calibrations and laboratory QC checks; and the second opinion in data interpretation. Laboratories document internal review in appropriate checklist forms that are kept on file with the rest of the project documentation and sample data. The internal data review process is generally described in Laboratory QA Manual and detailed in appropriate SOPs.

When conducted as a three-tiered process (Tier 1—Technical Review, Tier 2—Peer or Supervisor Review, and Tier 3—QA Review), internal review is an effective error-detection mechanism. Each tier in the review process involves the evaluation of data quality based on the results of QC data and on the professional judgment of the reviewer. The application of technical standards and professional experience to data review is essential in ensuring that data of known quality are generated consistently.

The use of LIMS makes internal review process rapid, reliable, and concurrent with data reporting. While Tier 1 reviewers typically work with hardcopy data, Tier 2 reviewers often access appropriate LIMS modules and review sample and QC data before final results are printed. They will also review the accompanying laboratory documentation that is available in hardcopy form only. Computerized confirmation of numeric acceptance criteria, such as the recoveries of laboratory QC check samples or surrogate standards, significantly reduces time and effort during review. Tier 3 Review is usually conducted on a hardcopy final data after analytical results and support documentation have been compiled into a single package.

Tier 1—Technical Review

Laboratory analysts who generate analytical data typically have the primary responsibility for their correctness and completeness. The analysts will review their work based on the QC criteria for each method and on project-specific requirements available from the SAP. This review will ensure that the following conditions have been observed:

- Initial and continuing calibrations have met the applicable QC requirements.
- Laboratory blanks are acceptable.
- Laboratory QC checks met the acceptance criteria.
- Project-specific requirements for sample preparation and analysis have been met.

Practical Tips: Seven steps of the laboratory process (continued)

5. Always specify preparation methods in the SAP.

6. Beware of a laboratory that reports sample results with more than two significant figures. By doing this, the laboratory may be misrepresenting the accuracy of measurements.

7. Remember that concentrations based on dry weight are higher than the wet weight concentrations.

8. Never make decisions based on preliminary data. Request that even expedited laboratory results undergo internal review prior to release.

9. Whenever practical, meet with laboratory Project Managers personally to establish a better relationship.

10. Notify the laboratory of your data package and EDD needs in advance and allow sufficient time for data package preparation. Three weeks should be sufficient for the preparation of data validation packages and the EDDs.

11. Include the cost of data packages and the EDDs in the project budget.

12. For proper budgeting, inquire whether the laboratory charges a fee for samples put on hold, kept in storage for long periods of time, and for sample disposal.

13. Be familiar with the laboratory policy on storing the unused portions of the samples. Request that the storage time be extended for the samples that may be analyzed for additional parameters at a later date.

4.4   ANALYTICAL TECHNIQUES AND THEIR APPLICATIONS

Environmental laboratories use a great variety of analytical methods for different types of organic and inorganic pollutant determinations. In this chapter, we will review the main instrumental techniques, their applications, and limitations in the analyses of environmental matrices, while focusing on qualitative aspects of environmental analysis.

4.4.1   Gas chromatography

Gas chromatography is arguably the most widely used analytical technique in the analysis of environmental pollutants. It is based on the ability of organic compounds to partition at a different speed between two different phases, mobile and stationary, in a chromatographic column. The technique of gas chromatography (or gas–liquid chromatography) owes its name to the mobile phase used for the separation, which is an inert gas (helium or nitrogen). As the mobile phase carries chemicals through the chromatographic column, partitioning takes place between the gaseous mobile phase and the stationary liquid phase inside the column.

A gas chromatograph has four major elements that perform the separation and the detection of separated compounds: an injection port, a chromatographic column inside an oven, a detector, and a data acquisition computer that controls the instrument operation, acquires data in a digital form, and stores them on electronic media.

Chromatographic column is the heart of a gas chromatograph. In the past, packed columns were used for all applications; these were glass or stainless steel tubes, 1–2 m long and about 5 mm in the outer diameter, filled with an inert material coated with a stationary phase. In the last 20 years, packed columns became for the most part obsolete in the environmental laboratory work and have been replaced with superior capillary columns. (They are still used for some applications, for example, atmospheric gas analysis.)

A modern gas chromatography capillary column is a small diameter tube made of fused silica glass with the walls coated with a film of a stationary phase. These flexible columns with the lengths ranging from 15 to 105 m and the internal diameters of 0.1– 0.75 mm are rolled into coils for mounting into the oven of a gas chromatograph.

Different stationary phases have different polarity and chemical composition and are designed for the separation of different classes of organic compounds. The proper selection of the stationary phase is a key to a successful separation. The selection process has been made easy due to the efforts of column manufacturers who conduct research in the area of environmental pollutant analysis and offer a range of columns specifically designed for different applications. The EPA validates analytical methods using these state-of-the-art columns and recommends them as part of the approved methods.

Chemicals are introduced into the column through the injection port at the head of the column. Two sample introduction techniques are predominant in environmental analysis:

- VOCs are purged from soil and water samples with a flow of an inert gas and concentrated in a trap kept at ambient temperature. After the purging has been completed, the trap is heated, and a reverse flow of inert gas flushes the VOCs from the trap onto the top of the column.
- SVOCs are extracted from environmental samples with organic solvents; the extracts are then concentrated to a 0.5–10 ml volume. A small volume (1–5 µl) is introduced into the injection port with a microsyringe; the high temperature at which the injection port is maintained instantly vaporizes the extract without destroying the chemicals in it.

As the mobile phase (inert gas) carries the chemicals through the column, repeated partitioning of molecules takes place between the mobile and the stationary phases, and molecules of same (or similar) chemical and physical nature are separated into groups. These groups reach the end of the column at different times. This process is called *elution*, and it can be compared to a marathon race: all runners start at the same time, but reach the finish line at different times due to the differences in their athletic ability.

At the end of the column is a detector that senses the presence of individual compound groups as a change in the analog electrical signal. The signal is amplified, digitized, sent to the data acquisition computer, and recorded as a chromatographic peak. The area under the peak (for some detectors, the height of the peak) is

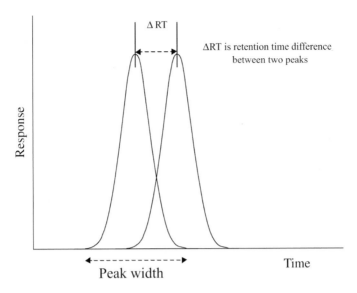

Figure 4.5 Chromatographic peak overlap and resolution.

proportional to the number of molecules of the detected compound. To quantify the amount, the area under the peak is compared to the area of a standard with a known concentration of each compound.

The time each group of molecules spends in the column from the moment they are introduced onto the top of the column to the moment they are recorded by the detector is called *retention time (RT)*. To speed up the partition process and reduce the total time needed for the separation of individual compounds, the oven may be gradually heated to as high as 400°C.

For accurate quantitation, chromatographic peaks must be sufficiently *resolved*, i.e. the two adjacent peaks must have as little overlapping between them as possible. Resolution is defined as the distance between two adjacent peak centers (ΔRT) divided by the average width of the peak, and it is measured in the units of time (Willard, 1988). The concepts of peak overlap and resolution are illustrated in Figure 4.5.

The quality of chromatographic separation depends on the composition of the stationary phase of the column and the instrument settings. The temperature regime of the oven; the flow rate of the mobile phase gas through the column; the temperature of the injection port—all of these factors influence compound retention time and peak resolution. The reduction in the chromatographic analysis time may adversely affect compound resolution. Because commercial laboratories always balance a need for a sufficient resolution with a need to perform analysis within the shortest possible period of time, the quality of chromatographic resolution is often traded for the speed of analysis.

The concepts of retention time and peak resolution form the very basis of chromatographic analysis. They also are the technique's worst limitations. The qualitative and quantitative uncertainty of chromatographic determinations originates from the following factors:

- Every chromatographic method identifies analytes based on their retention time by comparing the target analyte retention time in the chromatogram of a sample to that of a standard.
- The accuracy of compound quantitation depends on the quality of peak resolution.

Particularly unreliable are chromatography results for complex environmental matrices where the retention time and peak resolution of individual target analytes are affected by matrix interferences from dozens of non-target compounds also present in the sample. For such samples, chromatographic identification and quantitation becomes questionable for the following reasons:

- *False negative results*—Retention time shifts induced by chemical interactions in the column or by the changes in ambient temperature may cause a failure to identify target analytes.
- *False positive results*—Many organic compounds have similar chromatographic behavior and may have the same retention time. When identified by the retention time only, non-target analytes are often mistaken for target analytes.
- *Results with high bias*—Coelution with other target or non-target analytes affects the quantitation of the target analytes due to the degradation of chromatographic peak shape.

Recognizing the fact that shifts in retention time of individual compounds may cause false negative results, laboratories use *retention time windows* for target analyte identification. Retention time windows are experimentally determined retention time ranges for each target analyte. To minimize the risk of false positive results, EPA methods require that chromatography analysis be performed on two columns with dissimilar polarities. This technique, called *second column confirmation*, is described in Chapter 4.4.3. It reduces the risk of false positive results, but does not eliminate them completely.

Another technique that allows reducing the risk of false positive results is the use of *selective detectors*. These detectors are designed to distinguish between compounds of different chemical classes and identify them based on the presence of a specific functional group or a chemical element.

4.4.1.1 Detector selectivity and sensitivity

Gas chromatography detectors fall into three general categories based on their selectivity and sensitivity to various organic compounds:

1. *Non-selective*

   Detectors that respond to all classes of chemical compounds and that cannot discriminate between compounds of different chemical classes are called non-selective. The most widely used non-selective chromatographic detector is the FID, capable of detecting practically all of the organic compounds.

2. *Selective*

   Detectors that respond only to a certain class of chemical compounds have the property of selectivity. Selective detectors are designed to be particularly sensitive to specific groups of chemical compounds and not to be sensitive to all other compounds. Although selective detectors are intended to definitively identify specific classes of compounds, their selectivity may not be perfect.

3. Universal

These are detectors that respond to all classes of chemical compounds and at the same time provide definitive compound identification. The only chromatography detector that is universal, i.e. non-selective and selective at the same time, is a mass spectrometer.

Detector sensitivity is best explained in terms of *signal to noise ratio*, which is the minimum detectable quantity with a signal to noise ratio of two (Willard, 1988). Detector sensitivity is linked to the method detection limit, a concept that we routinely use in environmental project work. (The definitions of detection limits in environmental pollutant analysis are discussed in Chapter 4.5.1.) The MDLs, however, while being related to detector sensitivity, greatly depend on the analytical method, sample matrix, and the analyte itself. In this chapter, we will address detector sensitivity in relative terms by comparing sensitivities of various chromatography detectors.

Another important characteristic of a detector is the *linear dynamic range*: a range of analyte concentrations that produce a calibration line best described with a linear equation. Non-linear calibrations with subtle curvatures may be also used in the linear dynamic range of the detector.

The width of the linear dynamic range determines the magnitude of dilutions needed for accurate sample analysis and affects each analyte's PQL.

Understanding the issues of chromatographic detector selectivity, sensitivity, and linearity is important in the selection of analytical methods during project planning phase of the data collection process. Understanding the advantages and limitations of each detector is important for data interpretation conducted in the assessment phase. To further the knowledge of the advantages and limitations of detector applications, this chapter addresses the characteristics of the following common chromatography detectors:

- *Flame ionization detector*: non-selective, responds to virtually all organic compounds with the carbon-hydrogen bonds
- *Photoionization detector*: selective to aromatic and unsaturated compounds
- *Electrolytic conductivity detector (ELCD)* and *electron capture detector (ECD)*: selective to chlorinated and oxygenated compounds
- *Nitrogen–phosphorus detector (NPD)*: selective to compounds containing nitrogen and phosphorus
- *Flame photometric detector (FPD)*: selective to compounds containing sulfur and phosphorus
- *Mass spectrometer*: universal

Table 4.3 summarizes the applications these chromatography detectors for environmental pollutant analysis and their limitations that may affect the identification and quantitation of pollutants.

Table 4.3　Common chromatography detectors and their applications

Selectivity	Analytical Methods	Limitations and Interferences
Gas chromatography—analysis of compounds that can be volatilized without being decomposed or chemically rearranged		
	Flame ionization detector	
Almost all classes of compounds; non-selective	Non-halogenated volatile compounds (EPA 8015) Phenols (EPA 8041) PAHs (EPA 8100)	—All other chemical compounds present in the sample will interfere with the target analytes. —TPH analyses are affected by naturally occurring organic compounds in soils with high humic substance content.
	Photoionization detector	
Moderately selective to aromatic compounds when a lamp with ionization energy of 10.2 eV is used	Volatile aromatic hydrocarbons (EPA 8021)	—Many non-aromatic volatile organic compounds have a response on a 10.2 eV lamp. —Higher energy lamps (11.7 eV) ionize a wide range of volatile compounds. —Water vapor suppresses response.
	Electrolytic conductivity detector	
Highly selective to chlorinated compounds	Volatile chlorinated hydrocarbons (EPA 8021)	Sulfur and nitrogen compounds may interfere with target analytes.
	Electron capture detector	
Highly selective to halogenated and oxygenated compounds	EDB, DBCP (EPA 8011) Acrylamide (EPA 8032) Phenols (EPA 8041) Phthalates (EPA 8061) Organochlorine pesticides (EPA 8081) PCBs (EPA 8082) Nitroaromatics and cyclic ketones (EPA 8091) Haloethers (EPA 8111) Chlorinated herbicides (EPA 8151) CLP SOW for organic analysis	Interferences from: • Elemental sulfur (S_8) • Waxes, lipids, other high molecular weight compounds • Phthalate esters, which are common laboratory contaminants • Oil in PCB analysis
	Nitrogen–phosphorus detector	
Highly selective to compounds containing nitrogen and phosphorus	Organophosphorus pesticides (EPA 8141) Acrylonitrile (EPA 8031) Acetonitrile (EPA 8033) Nitrosamines (EPA 8070)	Hydrocarbons, fats, oils, waxes may interfere with organophosphorus pesticides.

Table 4.3 Common chromatography detectors and their application (Continued)

Selectivity	Analytical Methods	Limitations and Interferences
	Flame photometric detector	
Selective to sulfur and phosphorus compounds; more selective than NPD to compounds containing phosphorus	Organophosphorus pesticides (EPA 8141)	Sulfur compounds may interfere with organophosphorus pesticides.

High performance liquid chromatography—analysis of semivolatile and nonvolatile chemicals or analytes that decompose upon heating

	Ultraviolet/visible photometer	
All compounds with UV or visible (VIS) range absorbance	PAHs (EPA 8310) Nitroaromatics, nitramines (EPA 8330) Nitroglycerine (EPA 8332)	—Detection based on a single wavelength is prone to interferences. —Does not have the selectivity for positive compound identification.
	Diode array detector	
All compounds with UV or VIS absorbance	PAHs (EPA 8310) Nitroaromatics, nitramines (EPA 8330) Nitroglycerine (EPA 8332)	Interferences from: • Oily matrices • Trace level laboratory artifacts
	Fluorometer	
Highly selective to compounds with fluorescent emission spectra	PAHs (EPA 8310) *N*-methyl carbamates (EPA 8318)	Interferences from: • Alkyl amines • Phenols • Plant pigments • Oily matrices

Gas chromatography and high performance liquid chromatography

	Mass spectrometer	
All classes of organic compounds; each compound identified by its mass spectrum	*Gas chromatography:* VOCs (EPA 8260) and SVOCs (EPA 8270); dioxins/furans (EPA 8280, EPA 8290); CLP SOW for organic analysis	Interferences from: • Oily matrices • Trace level laboratory artifacts
	Liquid chromatography: solvent-extratable non-volatile compounds (EPA 8321, EPA 8325)	Interferences from: • Organic acids • Phenols • Trace level laboratory artifacts

EDB denotes ethylene dibromide.
DBCP denotes dibromochloropropane.

Flame Ionization Detector

The FID is a non-selective detector widely used in environmental pollutant analysis. It detects the carbon–hydrogen bonds, which are present in almost every organic compound. The FID has no response to compounds that do not contain carbon–hydrogen bonds (carbon disulfide, carbon tetrachloride) and is insensitive to moisture and permanent atmospheric gases. The FID has a wide linear dynamic range and is rugged and dependable.

The FID earns the name from a high temperature flame fueled with a flow of hydrogen and air that burns inside the detector body. Two electrodes detect the background electrical current from ions and electrons produced in the flame. Organic compounds that elute from the column are burned in the flame and create an influx of new ions and electrons, elevating the background current. The change in the current is proportional to the number of burned molecules.

The usefulness of the FID for individual compound analysis is limited by its lack of selectivity. Almost every organic compound will produce an FID response that is proportional to the number of $-CH_2-$ groups in the organic molecule; the intensity of the response varies for functional groups of different types. The best FID application is petroleum hydrocarbon fuel analysis. In TPH analyses, fuels are identified by comparing a chromatographic pattern (fingerprint) of a sample to patterns of petroleum fuel standards. To determine the best match, standards of several petroleum fuels (for example, gasoline, diesel, and motor oil shown in Figure 2.5) are analyzed. The quantitation is based on the combined area of all peaks within a certain carbon range representing a fuel concentration in the sample.

In individual compound analysis with the FID, analyte identification relies *solely* on retention time. As a result, single compound determinations (for example, individual phenols or PAHs) are usually gravely affected by interferences that often render the results unreliable or unusable. Because the FID methods for individual compound analysis are so susceptible to false positive results, they should be used with caution and only for interference-free matrices.

Photoionization Detector

Volatile aromatic and chlorinated compounds are usually analyzed with the photoionization detector/electrolytic conductivity detector (PID/ELCD) combination in EPA Method 8021. In this method, the PID detects aromatic compounds, typically the volatile constituents of petroleum fuels (BTEX) and oxygenated additives, and the ELCD detects chlorinated solvents. Both detectors are considered to be selective for the target analytes of EPA Method 8021. But are they sufficiently selective for making unambiguous decisions on the presence and the concentrations of these analytes?

The PID uses UV radiation to ionize the VOC molecules; the resultant increase in ion current is measured as a response to the amount of these ions. The source of UV radiation is a lamp with the ionization potential that may range from 8.4 to 11.7 eV. Compounds with ionization potentials lower than the lamp's ionizing energy will produce a response on the PID. The selectivity of the PID is in inverse correlation to the energy of the lamp: the higher the energy of the lamp, the less selective the detector, as illustrated in Example 4.9.

Example 4.9:   The PID selectivity

- Applications of the hand-held field PID were discussed in Chapter 3.9.2. Because a field PID is typically equipped with a high-energy lamp (10.6 or 11.7 eV), it becomes a non-selective detector capable of responding to a great number of VOCs.
- Laboratory PIDs have a better selectivity due to the lower energy lamps (typically 10.2 eV) used in laboratory instruments. In addition to aromatic compounds, the laboratory PIDs respond to unsaturated and oxygenated compounds, such as cyclohexene, trichloroethylene (TCE), methyl tertiary butyl ether (MTBE), and many others.
- If BTEX components and MTBE are not sufficiently resolved in a complex gasoline mixture, their identification and quantitation with a PID becomes questionable. False positive results and high bias are common in gasoline component analysis with EPA Method 8021.

The PID, while not having the ultimate selectivity for aromatic volatile compounds, nevertheless exhibits the best sensitivity to these compounds relative to other detectors and has a wide linear dynamic range.

Electrolytic Conductivity Detector

The ELCD is a detector that is highly sensitive and selective to chlorinated compounds, volatile and non-volatile alike. Chlorinated compounds eluting from the column are pyrolized in a flow of hydrogen inside a nickel tube furnace at 850°C and produce hydrochloric acid. The acid is transported into an electrolytic conductivity cell; the increase in the cell's conductivity is proportional to the number of pyrolized molecules. Compounds containing organic nitrogen and sulfur will undergo the same reaction to produce ammonia and hydrogen sulfide; such interferences are very rare in VOC analysis of environmental matrices.

Compound identification with the ELCD, although primarily relying on retention time, is dependable, and quantitation is accurate provided that sufficient peak resolution has been achieved. The ELCD has a relatively narrow linear dynamic range and an unsurpassed sensitivity to chlorinated solvents, detecting the concentrations in the part per trillion range in water samples.

Electron Capture Detector

The ECD is a selective detector used in the analysis of halogenated compounds (pesticides and herbicides) and oxygenated compounds (phthalates and ketones). The ECD detection mechanism is based on the depletion of the background signal between two electrodes in the presence of compounds that absorb electrons. One electrode is coated with a radioactive source (nickel-63 foil) that emits a flow of beta particles (high-energy electrons). Beta particles bombard the carrier gas and produce a cloud of ions and free electrons inside the ECD cell. When electron-absorbing compounds enter the ECD cell, they capture the free electrons and combine with them, creating a decrease in the background signal. The intensity of the signal is proportional to the number of electron-absorbing groups: a molecule with six chlorines will have a greater response than a molecule with only one chlorine. That is why in PCB analysis, the response to Aroclor 1260 (60 percent chlorination by weight) is greater than the response to Aroclor 1016 (16 percent chlorination). The ECD has a relatively narrow linear dynamic range and a superior sensitivity to chlorinated compounds detecting the concentrations in the part per trillion range.

Although compound recognition in the ECD methods is based on retention time, superb selectivity of this detector makes compound identification and quantitation reliable in interference-free matrices. The ECD's supreme sensitivity to halogens and oxygen is also its greatest weakness. Example 4.10 illustrates this point.

Example 4.10: Interferences in ECD analysis

The ECD is used in analysis of chlorinated pesticides (EPA Method 8081), herbicides (EPA Method 8151), and PCBs (EPA Method 8082). These analyses are prone to interferences from naturally occurring and manmade non-target compounds, such as the following:

- Naturally occurring oxygenated compounds are omnipresent in environmental matrices and to a varying degree affect virtually every sample.
- Another formidable type of naturally occurring interference is elemental sulfur that may be present in soil.
- Among manmade interferences are the phthalate esters, the oxygenated compounds that are ubiquitous laboratory contaminants.
- Sporadic trace contamination originating from inadequate glassware washing at the laboratory also interferes with analysis.

Applying various cleanup techniques to remove naturally occurring compounds and taking measures to eliminate trace level contamination in the glassware can considerably reduce the detrimental effects of these interferences on analysis of pesticides, herbicides, and PCBs. Even so, they are a major source of elevated detection limits and false positive results in environmental matrices.

Nitrogen-Phosphorus Detector

The NPD, also known as *thermionic emission detector*, is similar in design and detection mechanism to the FID. It is a flame ionization detector with a low temperature flame and a rubidium ceramic flame tip (also called *a bead*) that gives the NPD its selectivity to compounds containing nitrogen and phosphorus. As analytes elute from the column, they undergo a surface catalytic reaction with the bead, and the resulting ions are detected as an increase in the background signal. The low temperature flame suppresses the ionization of compounds that do not contain nitrogen or phosphorus. By adjusting the settings, the NPD can be made selective only to phosphorus compounds.

Flame Photometric Detector

The FPD is a special flame emission photometer used for the detection of phosphorus and sulfur compounds. Compounds eluting from the chromatographic column are burned inside the detector body in a low temperature flame where phosphorus and sulfur form the species with characteristic emission spectra detected with a photometer. The FPD is more sensitive and more selective to phosphorus than the

NPD; it is also less common in environmental laboratories. Either detector has a moderate linear dynamic range.

The NPD and FPD are used in EPA Method 8141 for analysis of organophosphorus pesticides, which may contain chlorine in addition to phosphorus, making their detection also possible with the ECD.

Mass Spectrometer

Mass spectrometer is the only detector used in chromatographic analysis that is capable of definitively detecting and identifying any organic compound. The supremacy of mass spectrometer made it a sought after detector in other analytical techniques, such as liquid chromatography and elemental analysis. In gas chromatography analysis, a mass spectrometer is used in EPA Methods 8260 (VOCs), 8270 (SVOCs), 8280 (low resolution dioxins/furans) and 8290 (high resolution dioxins/furans). Samples collected for CWA, NPDES, and SDWA dioxins/furan monitoring are analyzed with a GC/MS EPA Method 1613. The CLP SOW includes VOC, SVOC, and dioxin analyses with GC/MS methods.

A low resolution mass spectrometer used for routine analysis by EPA Methods 8260 and 8270 consists of three main parts: an ion source, a mass analyzer (both kept under vacuum conditions), and a detector. Although a variety of GC/MS instruments with different designs are available, the most common in the environmental laboratory are those with electron impact ion sources and quadrupole mass analyzers as they are best suited for the coupling with gas chromatography columns. These stable and reliable detectors with a relatively broad linear dynamic range became the workhorses of the environmental laboratory.

In the electron impact mass spectrometer, analyte molecules bombarded in the ion source with a flow of electrons, produce charged fragments. The negatively charged ions are deflected and removed from the system, whereas the positively charged fragments are accelerated and enter the quadrupole mass analyzer. The fragments are separated by mass and velocity under the effects of a direct electrical current and a radio frequency applied to the quadrupoles. Only the ions with a single charge (mass to charge ratio m/z is equivalent to mass) are retained in the mass analyzer. Each group of ions with the same m/z value has its own trajectory as it reaches the detector. As ion beams of different m/z values enter the electron multiplier of the detector, they are recorded as a change in the electrical signal. Although the mass analyzer scans a certain range of ion masses, it is programmed to record only the ones that are known to be important for the make up of a target analyte mass spectrum.

Every organic chemical has a *mass spectrum*, which is a combination of ions with different masses and different intensities (abundances). To identify a compound, its mass spectrum is compared to the mass spectra of standards, analyzed under the same instrument settings, and to the EPA/National Institute for Standards and Technology (NIST) mass spectra library. The EPA/NIST library is stored in the database of the computer that operates the instrument. A comparison to the library spectra is possible only if there is consistency in the compound spectra generated by different GC/MS systems at hundreds of environmental laboratories. To achieve such consistency, the EPA methods for GC/MS analysis include the *mass spectrometer tuning* requirements, which assure that the spectra, no matter where they are obtained, are essentially similar. The tuning is verified by analyzing standards, such as bromofluorobenzene (BFB) for VOC analyses and decafluoro-triphenylphosphine (DFTPP) for SVOC analyses, and by verifying that their spectra meet the method's specifications.

Compound identification in GC/MS methods is based on the retention time and the mass spectra interpretation; the quantitation is based on the abundance of a primary (characteristic) ion. For a compound to be positively identified, all of the ions in the spectrum must be detected at one and the same retention time, which corresponds to the retention time of this compound in the calibration standard. The retention times of the primary ion and one or two secondary ions are typically monitored for this purpose. A combination of retention time and mass spectra is a formidable compound identification tool. That is why compound identification mistakes are relatively rare in GC/MS analyses, and they usually occur due to analysts' lack of experience.

The capability to identify compounds outside the target analyte list is one of the greatest advantages offered by GC/MS. Searching the EPA/NIST spectral library, the computer finds the best matching library spectra for an unknown compound and tentatively identifies it using the best fitting spectrum. The retention time information is not available for this compound as it is not part of the calibration standard, hence only tentative identification. These tentatively identified compounds (TICs) can be also quantified to provide an estimated concentration value.

The low resolution mass spectrometers used in EPA Methods 8260 and 8270 are not as sensitive as some of the selective chromatographic detectors (for example, the ECD) and for this reason are not capable of reaching the low detection limits that may be required for some DQOs. The mass spectrometer scans a large number of ion masses in a short period of time (for example, in EPA Method 8270, a mass range of 35–500 is scanned in 1 second) and dwells only briefly on each detected mass. In such full scan mode, the sensitivity of detection is traded for a wide range of detected ions. It is also affected by the background spectra (an equivalent of the electrical signal noise).

To improve sensitivity, selected ion monitoring (SIM) mode may be used for the detection of routine VOCs and SVOCs and in low resolution dioxin/furan analysis (EPA Method 8280). In the SIM mode, only specific ions from the analyte's spectrum are scanned for; the detector's dwelling time on each ion is increased resulting in higher sensitivity. A mass spectrometer operated in the SIM mode is approximately ten times as sensitive as one in the full scan mode. The SIM mode has limitations, such as the capacity to monitor only a limited number of ions and the need to monitor multiple ions for each compound to improve the degree of confidence in compound identification. That is why typically no more than 20 compounds can be analyzed simultaneously in the SIM mode.

High resolution mass spectrometers used in dioxin/furan analyses (EPA Method 8290 and EPA Method 1613) offer supreme sensitivity, down to parts per quadrillion concentration levels in purified samples. These are research grade instruments with a principle of operation that is based on the separation of ions in a magnetic field followed by ion optics detection. Dioxin/furan analysis is a unique application performed at specialty laboratories, which have the proper resources and personnel to own, maintain, and operate high resolution GC/MS instruments.

4.4.2 Liquid chromatography

High performance liquid chromatography is a separation technique used for analysis of a wide range of semivolatile and non-volatile chemical pollutants in environmental samples. It is particularly valuable for analysis of thermally unstable compounds that decompose upon heating in a gas chromatography system and for analysis of large molecules that cannot be volatilized in the injection port of a gas chromatograph. Liquid chromatography gets its name from the liquid mobile phase, which is an organic solvent, water, or the mixture of two. Because the solvents are delivered under pressure, the technique was originally called high pressure liquid chromatography, but today it is commonly referred to as high performance liquid chromatography or HPLC.

Similar to GC instruments, HPLC instruments consist of an injection port, a separation column, a detector, and an instrument control/data acquisition computer. The use of liquid as a mobile phase influenced the design and construction materials of HPLC instrumentation elements. A sample extract or an aqueous sample is introduced into the separation column through an injection loop that can be programmed to receive various volumes of liquid (5 μl to 5 ml).

Compared to GC columns, HPLC columns are short and thick, ranging from 10 to 25 cm in length and 2 to 4.6 mm in internal diameter. They are filled with an inert material (silica, polymer resin), which is coated with a stationary phase. In *normal phase* HPLC, the mobile phase is *less* polar than the stationary phase. In *reverse phase* HPLC, the opposite is true, and the mobile phase is *more* polar than the stationary one. Reverse phase HPLC is the technique of choice for environmental applications. Similar to GC columns, analyte-specific HPLC columns are recommended in the published methods.

For liquid chromatography separation to take place, samples or sample extracts must be miscible with the mobile phase. The separation is possible due to specific interactions between the analyte molecules and the mobile and stationary phases. To decrease the time of analysis, gradient elution is applied, with the polarity of the mobile phase solvent increasing through analysis time. Up to four solvents of different polarities may be mixed to achieve the fastest and the most complete separation.

Various flow cell detectors with the spectral detection mechanism may be part of HPLC analytical systems, such as the following:

- *Ultraviolet/visible (UV/VIS) photometer*—a non-selective detector that responds to a majority of organic compounds
- *Fluorometer*—a detector selective to compounds with fluorescent emission spectra
- *Diode array UV/VIS detector*—a universal detector that responds to a majority of organic compounds and provides a definitive identification
- *Mass spectrometer*—a universal detector

Table 4.3 summarizes the detectors used in common HPLC methods and their applications and limitations.

Ultraviolet/Visible Photometer
The UV/VIS photometer uses the UV/VIS absorption of organic compounds as the detection mechanism. The UV/VIS region spans the wavelength range from 190 to 800 nonameters (nm). Most organic compounds absorb light in this region to produce a distinctive UV/VIS spectrum.

The UV/VIS photometer has a radiation source and a flow cell with two channels. The sample channel measures the signal generated by UV/VIS absorption of the sample, and the reference channel compensates for the system's background absorbance. After the analytes have been separated in the column, they pass through the sample channel. The increase in absorbance at a characteristic UV or VIS wavelength is proportional to the analyte concentration. This type of a detection mechanism is possible in the absence of absorption from the mobile phase (organic solvents and water), and the detection is most sensitive at the wavelength in the analyte's UV/VIS spectra that has the highest absorbance.

The UV/VIS detector has a broad linear dynamic range. However, as illustrated in Example 4.11, it does not offer the selectivity required for analysis of complex environmental matrices. Data obtained from UV/VIS detectors should be confirmed by other methods. Neither does the UV/VIS detector offer the sensitivity required for many DQOs, since a selected fixed wavelength does not necessarily have the maximum absorbance for all target analytes.

Example 4.11:   Selectivity of the UV/VIS detector

Fixed wavelength UV/VIS detectors have poor selectivity because a great number of organic compounds may have a response at the same wavelength making an absorption band from a single wavelength inadequate for positive compound identification. For example, a UV/VIS detector with the wavelength of 254 nm is commonly used in analysis of environmental samples. Different classes of compounds, such as PAHs (EPA Method 8310) and trace explosives (EPA Method 8330), are detected at this wavelength. All of the compounds on the analyte lists for these two methods will have a response at the wavelength of 254 nm, and so will many non-target organic compounds that may be present in samples. Relying on the single wavelength UV/VIS detection for compound identification often produces false positive results; only the full UV/VIS spectra provides reliable compound identification.

Fluorometer
The fluorometer or fluorescent detector is used for analysis of PAHs in environmental matrices. When brought into an excitation state with radiation of a certain wavelength from the UV/VIS region, PAHs, which are conjugated aromatic compounds, produce fluorescent emission spectra. In fluorometry analysis, a characteristic emission wavelength is selected from the complete emission spectra of each individual PAH. Compounds may be excited with the same wavelength; however, the selected emission wavelengths must be different. The signal from the emission band is proportional to the concentration of the PAH compound; it is recorded and serves as the basis for quantitation. A variable wavelength fluorometer is programmed to switch the excitation/emission wavelength pairs according to the retention times of individual PAH compounds. As the

compounds elute from the column and enter the detector flow cell, the excitation/ emission pairs change to achieve the maximum sensitivity and selectivity for each target analyte.

The fluorometer offers superb sensitivity and selectivity to PAHs, although its linear dynamic range is relatively narrow. The MDLs of EPA Method 8310 are by an order of magnitude lower than their EPA Method 8270 counterparts.

Although an excellent detector for PAHs, the fluorometer is not widely used in environmental analysis, as the number of environmental pollutants with fluorescent spectra is limited. The sensitivity and selectivity of the fluorometer are also used in the *N*-methyl carbamate pesticide analysis (EPA Method 8318). These compounds do not have the capacity to fluoresce; however, when appropriately derivatized (chemically altered), they can be detected fluorometrically. The process of derivatization takes place *after* analytes have been separated in the column and *before* they enter the detector. This technique, called *post column derivatization*, expands the range of applications for the otherwise limited use of the fluorometer.

Diode Array Detector

Reliable compound identification can be achieved with the scanning wavelength UV/ VIS detectors that utilize diode arrays for simultaneous measurements of the whole spectrum. A broad wavelength range can be measured with the diode array detector in a fraction of a second; the obtained data are sent to the data acquisition system where the spectrum is stored. It may be retrieved from the memory for a two- or three-dimensional presentation (e.g. absorbance *versus* wavelength and time). The laboratories optimize compound quantitation and the sensitivity of analysis by selecting from the full spectrum a characteristic wavelength with the highest absorbance. The diode array detector is a universal detector since it positively identifies compounds based on a comparison of their *full* spectra to a library of spectral standards.

Mass Spectrometer

Mass spectrometer can be used as a detector in HPLC analysis. A special interface between the column and the ion source removes the mobile phase solvent and vaporizes the nonvolatile analytes without destroying or rearranging them. The molecules are ionized, and the ions are directed into a quadrupole mass analyzer or into a magnetic sector spectrometer.

The HPLC/MS technique used in EPA Method 8321 is best suited for analysis of thermally unstable compounds that are hard to analyze with conventional GC methods, such as organophosphorus pesticides, chlorinated herbicides, and carbamates. In this technique, the detection with mass spectrometry provides the ultimate selectivity. The sensitivity for each individual compound depends on the interferences in a given environmental matrix and on the chemical nature of the analyte.

The HPLC/MS technique has been only recently initiated into environmental applications. It is considered a non-routine application, and is conducted mostly by the laboratories specializing in pesticide analysis.

4.4.2.1 Ion chromatography

Ion-exchange chromatography is a liquid chromatography application that allows the separation of cationic and anionic species. Ion chromatography, a technique used for the separation of dissolved inorganic ions, is employed in EPA Methods 9056 and 300.1 for analysis of anions (fluoride, chloride, bromide, nitrite, nitrate, ortho-phosphate, sulfate) and inorganic water disinfection by-products (bromate, chlorate, chlorite). Perchlorate, a toxic component of solid rocket propellant, is also analyzed with ion chromatography in the EPA Methods 314.1 and 9058. Ion chromatography has been successfully used in EPA Methods 218.6 and 7199 for the detection of hexavalent chromium in aqueous samples at concentrations as low as $0.3\,\mu g/l$.

To achieve the separation of anions, the sample is eluted through three different ion exchange columns into an electrical conductivity detector. The first two columns, a pre-column (also called a guard column) and a separating column, are packed with a strongly basic anion exchange resin. Ions are separated into discrete groups based on their affinity for the exchange sites of the resin. The third column is a suppressor column that reduces the background conductivity of the mobile phase (sodium carbonate in water) to a negligible level and converts the anions into their corresponding acids. The concentrations of separated anions in their acid form are then measured with an electrical conductivity cell. Anions are identified based on a comparison of their retention times to these in the calibration standards. Measuring the peak height (area) and comparing it to a calibration curve generated from standards allows compound quantitation. In ion chromatography analysis, compound confirmation is not necessary due to high specificity of the separation column and of the detector.

4.4.3 Compound confirmation in chromatography methods

Individual compound identification in all GC methods with the exception of GC/MS relies on the compound retention time and the response from a selective or non-selective detector. There is always a degree of uncertainty in a compound's identity and quantity, particularly when non-selective detectors are used or when the sample matrix contains interfering chemicals. To reduce this uncertainty, confirmation with a second column or a second detector is necessary. Analyses conducted with universal detectors (mass spectrometer or diode array) do not require confirmation, as they provide highly reliable compound identification.

Qualitative confirmation

Qualitative compound confirmation may be performed either with a second column or with a second detector. *Second column confirmation* technique consists of analyzing the sample on two columns with dissimilar polarities. Each column is calibrated with the same standards, and the same calibration acceptance criteria are applied. For the presence of a compound to be confirmed, its retention time values obtained from each column must fall into respective retention time windows. If a peak falls within the retention time window on one column, but not on the second column, the compound is not considered confirmed and should not be reported.

Second column confirmation is an imperfect technique, prone to false positive detection, particularly if non-selective or low selectivity detectors are used. Even the most selective detectors may not be fully capable of correctly identifying target analytes in complex environmental matrices as illustrated in Example 4.12.

Example 4.12:   Second column confirmation

Second column confirmation must be used in pesticide, PCBs, and chlorinated herbicide analyses by EPA Methods 8081, 8082, and 8151, respectively. In these methods, two columns with dissimilar polarities and two ECDs provide compound identification and quantitation. This technique produces a lower rate of false positive results, but does not eliminate them completely. This is particularly true for low concentrations of pesticides and herbicides, where non-target compounds, such as constituents of the sample matrix or laboratory artifacts listed in Table 4.3, produce chromatographic peaks on both columns. These interference peaks cannot be distinguished from the target analytes based on retention time only and cause false positive results.

Second detector confirmation is another compound confirmation technique. Two detectors with selectivity to different functional groups are connected in series to one column or in parallel to two columns. For example, two detectors, a UV/VIS detector and a fluorometer, connected in series to the HPLC column are used in the EPA Method 8310 for analysis of PAH compounds. If the second detector is connected to a second column of a dissimilar polarity, then the confirmation becomes even more reliable. An example of such a configuration is organophosphorus pesticides analysis: the samples may be initially analyzed with an NPD, and then confirmed on a different column with an ECD or a FPD.

The mass spectrometer provides the most reliable compound identification in chromatography methods. When the concentration levels are sufficiently high and GC/MS or HPLC/MS instruments are available, they should be the preferred confirmation techniques. However, not every detection may be confirmed by the mass spectrometry methods due to the limited sensitivity of the mass spectrometer.

Quantitative confirmation

The confirmation of a compound's identity is only one half of the overall confirmation procedure; quantitative confirmation is the other half. Compound concentrations calculated from analyses on two columns or two detectors must be in agreement. The EPA recommends a 40 percent difference (calculated as the RPD shown in Equation 1, Table 2.2) as a threshold value for making decisions on the presence or absence of a compound (EPA, 1996a). This means that the concentrations obtained from two columns or two detectors that agree within 40 percent indicate the presence of an analyte, provided that the retention time confirmation criterion has been also met.

If the RPD exceeds 40 percent, confirmation becomes questionable, especially if one result is significantly higher (orders of magnitude) than the other one. For such situations, laboratories develop internal policies and rules for data reporting. Because of the liability associated with data reporting, laboratories typically avoid committing to a single concentration value and report both results. This places the burden of choosing between results on the data user.

The EPA provides further guidance for decision-making in this area: when the RPD between two results exceeds 40 percent, the EPA conservatively recommends selecting the higher concentration as a true one (EPA, 1996a). This practice, however, often leads to false positive results and may be the cause of unnecessary site remediation. From a practical perspective, in the absence of matrix interferences for an analyte to be present in the sample the agreement between the two results should be better than 40 percent. If matrix interferences are obvious, a chemist experienced in data interpretation should evaluate the chromatograms and make a decision on the presence or absence of the analyte in the sample.

The complexities of the second detector and second column confirmation are illustrated in Example 4.13.

Example 4.13:   Second detector/second column confirmation

Gasoline and its components are usually analyzed with the PID/FID combination (EPA Methods 8021 and 8015). In these methods, the PID and the FID may be connected in series to one column or in parallel to two columns. The PID is used for the analysis of individual compounds (BTEX and oxygenated additives, such as MTBE), and the FID is used for TPH analysis.

The non-selective FID does not provide the second detector confirmation for BTEX, whereas the PID has only a marginal selectivity to aromatic compounds. In a typical analysis of gasoline in water by EPA Method 8021, benzene and MTBE results tend to be biased high due to coelution with other constituents; false positive results are also common.

Confirmation with a second column of a dissimilar polarity and a second, selective detector is necessary for the correct identification and quantitation of BTEX and oxygenated additives. And yet, even when a dissimilar column and a second PID are used in confirmation analysis, false positive detection of MTBE often takes place.

Because gasoline contains dozens of individual chemicals, coelution of compounds is typical during routine laboratory analysis, giving rise either to false positive results, notably for benzene and MTBE, or to concentrations that are biased high. GC/MS analysis by EPA Method 8260 is the *only* reliable method of confirming the presence of benzene and MTBE and for obtaining their correct concentrations.

Second column and second detector confirmation are not always necessary. If the source of contamination is known and there is certainty that the contaminants are, in fact, present at the site, confirmation of compound identification may not be necessary. For example, the confirmation of pesticides in the samples collected at a pesticide manufacturing facility is not needed as the source of contamination is known. Confirmation of pesticide concentrations in this case is still necessary.

Petroleum fuel analysis does not require second column or second detector confirmation. Fuels are identified based on their fingerprints or characteristic patterns of multiple peaks similar to ones shown in Figure 2.5. Each peak represents an individual chemical constituent, and each fuel has a unique combination of these constituents forming a characteristic pattern or a fingerprint. The fingerprints obtained from the samples are compared to these of fuel standards to identify the best match and to determine the type of fuel contamination. Quantitation of petroleum products is done using the combined area of all peaks.

On the other hand, PCBs, which are also the mixtures of individual chemical constituents (congeners) and which also have characteristic and recognizable chromatographic patterns, require second column confirmation. This is because they are quantified using individual peaks, the presence of which must be confirmed on the second column.

Practical Tips: Chromatography methods

1. Give preference to GC/MS methods for site investigation projects since unknown sites will be better characterized with methods that are highly reliable in compound identification.

2. Choose less expensive GC methods for site remediation projects, if contaminants of concern are known.

3. On long-term monitoring projects, occasionally confirm results of GC analysis with GC/MS.

4. Always give preference to chromatography methods that employ selective detectors, for example, for PAHs choose EPA Methods 8310 and 8270 over EPA Method 8100; for phenols choose EPA Method 8270 over EPA Method 8041/FID.

5. Request tentative identification of unknown individual compound peaks with GC/MS, particularly when they are sizable or recurring. (Exceptions are tentatively identified compounds in petroleum fuels, which are fuel constituents, such as aliphatic and aromatic hydrocarbons.)

6. Request that the laboratory confirm all chromatography results with a second column or a second detector, when required by the method. (Laboratories do not automatically do this.)

7. Whenever appropriate for the types of analytes and the range of concentrations, request that confirmation be performed with GC/MS methods.

8. Beware of low concentration pesticide or herbicide results, particularly when many analytes are reported: these may be false positive results produced by trace level oxygenated organic compounds or by laboratory artifacts.

9. For low detection limit analysis of PAHs, give preference to EPA Method 8270/SIM over EPA Method 8310. The detection limits may be comparable, but analysis with EPA Method 8270/SIM is less affected by interferences from oily matrices.

4.4.4   Trace element analysis

Metal and metalloid pollutant analysis of environmental matrices may be conducted with several elemental analysis techniques. For analysis to be possible, all matrices, including unfiltered groundwater, leaching procedure extracts, industrial and organic wastes, soils, sludges, and sediments must be *digested* with acid. Exceptions are drinking water samples and groundwater samples that have been filtered and acidified for dissolved metal analysis.

For routine determinations of metals and metalloids, laboratories use the following common trace element analysis techniques:

- Inductively coupled plasma–atomic emission spectrometry (ICP–AES)
- Atomic absorption (AA) spectrophotometry
- Cold vapor atomic absorption (CVAA) spectrophotometry for mercury analysis

For special applications, laboratories offer the following non-routine analytical techniques:

- Inductively coupled plasma–mass spectrometry (ICP–MS)
- X-Ray fluorescence spectrometry
- Gaseous hydride AA spectrophotometry
- Cold vapor atomic fluorescence spectrophotometry for mercury analysis

Trace element analysis is less labor-intensive and time-consuming than analysis of organic compounds. Modern analytical instruments that use inductively coupled plasma (ICP) are capable of simultaneously determining a large number of elements in a matter of minutes. In fact, the state-of-the-art ICP–AES instruments became the workhorses of the environmental laboratories as they offer multielement detection with the selectivity and sensitivity in the part per billion range that meet the DQOs of most environmental projects. ICP–MS has even better selectivity and sensitivity (in the part per trillion range) and in the recent years has been gaining a firm footing at the laboratories specializing in drinking water analysis. The graphite furnace atomic absorption (GFAA) technique has the sensitivity and selectivity that equal those of ICP–MS.

Table 4.4 compares the sensitivities of various elemental analysis techniques based on instrument detection limits (IDLs).

When planning for elemental analysis methods and interpreting the data in the assessment phase of the project, we need to have a good understanding of the intended use of the data and of the limitations that the method and the matrix place on the data. Because elemental analysis techniques are susceptible to chemical and matrix interferences and have a range of sensitivities and varying degrees of selectivity, we need to be particularly careful in selecting the methods that are appropriate for the project DQOs. To direct us towards the selection of a proper method, the discussion presented in the following chapters focuses on main technical aspects of common elemental analysis techniques and summarizes their limitations.

4.4.4.1 Inductively coupled plasma methods
EPA Methods 6010 (ICP–AES), 200.7 (ICP–AES), and 6020 (ICP–MS) use inductively coupled argon plasma as the source of excitation to produce emission spectra of metals and metalloids. (Hence another term for this technique, the ICAP, or inductively coupled argon plasma.) In ICP–AES or ICP–MS instruments, a liquid sample is nebulized (turned into aerosol) with a flow of inert carrier gas

Table 4.4 Comparison of instrument detection limits for various elemental analysis techniques

Element	Instrument detection limit, µg/l					
	Radial ICP–AES[1]	Axial ICP–AES[2]	ICP–MS[1]	FLAA[1]	GFAA[1]	Special methods
Antimony	30	1.9	0.07	70	3	NA
Arsenic	50	2.6	0.025	NA	1	2[3]
Beryllium	0.3	1.4	0.025	5	0.2	NA
Cadmium	4	0.6	0.006	2	0.1	NA
Chromium	7	1.0	0.04	20	2	NA
Lead	40	2.6	0.005	50	1	NA
Mercury	NA	NA	NA	NA	NA	0.2[4]
						0.0005[5]
Selenium	75	4.9	0.093	NA	2	2[3]
Silver	7	0.50	0.003	10	0.2	NA
Thallium	40	5.0	0.03	100[6]	1[7]	NA

[1] APHA, 1998.
[2] Courtesy Agricultural and Priority Pollutants Laboratory, Fresno, California.
[3] Gaseous hydride; EPA, 1996a
[4] Cold vapor atomic absorption; EPA, 1996a.
[5] Cold vapor atomic fluorescence; EPA, 1998d.
[6] EPA, 1983.
NA denotes "Not analyzed with this method."

(argon) and is transported into the plasma torch. The torch gains its power from a radio frequency source and is fueled with argon gas producing the ionized argon toroidal or annular plasma with the temperatures ranging from 6000 to 8500 °C.

The tail of the plasma formed at the tip of the torch is the spectroscopic source, where the analyte atoms and their ions are thermally ionized and produce emission spectra. The spectra of various elements are detected either sequentially or simultaneously. The optical system of a sequential instrument consists of a single grating spectrometer with a scanning monochromator that provides the sequential detection of the emission spectra lines. Simultaneous optical systems use multichannel detectors and diode arrays that allow the monitoring of multiple emission lines. Sequential instruments have a greater wavelength selection, while simultaneous ones have a better sample throughput. The intensities of each element's characteristic spectral lines, which are proportional to the number of element's atoms, are recorded, and the concentrations are calculated with reference to a calibration standard.

In the ICP–MS instrument, a mass spectrometer is used as a detector. Ions produced in the plasma are introduced through an interface into a mass spectrometer, sorted according to their mass-to-charge ratios, and quantified.

The ICP–AES instruments may have different configurations with the torch, which may be positioned either horizontally (axial ICP) or vertically (radial ICP). Because of the longer sample residence time in the axial ICP torch, the axial ICP–AES instruments are more sensitive than the radial ones. The state-of-the-art axial ICP–AES instruments (also called the *trace ICP*) have the high sensitivity that is required for trace element analysis of drinking water.

A simultaneous ICP–AES is the instrument preferred by the laboratories because it provides a fast and sufficiently sensitive analysis of multiple elements. Mercury is the only exception, as the high detection limit of the ICP–AES analysis for mercury does not meet the regulatory requirements. That is why mercury is analyzed with a much more sensitive CVAA technique.

ICP–AES analysis is susceptible to spectral and non-spectral interferences; spectral interferences constitute the greatest limitation of this otherwise sensitive and practical technique.

Spectral interferences originate from background ions contributed by the plasma gas, reagents, and from overlapping spectral lines from the constituents of the sample matrix. High background interferences reduce the sensitivity of analysis, and the overlapping spectral lines cause false positive results. A well-documented example of interferences from the overlapping spectral lines is the false positive detection of arsenic and thallium in soil with high aluminum concentrations.

To reduce the detrimental effects of spectral interferences on element quantitation, laboratories select the spectral lines that are least affected by the background, and use the background compensation and interelement correction routines as part of the analytical procedure. The instrument software uses equations to compensate for overlapping spectral lines; the effectiveness of these equations in eliminating spectral interferences must be confirmed at the time of sample analysis. That is why laboratories analyze a ***daily interelement correction standard*** (a mixture of all elements at a concentration of 100 mg/l) to verify that the overlapping lines do not cause the detection of elements at concentrations above the MDLs.

The following ***non-spectral*** (physical and chemical) interferences from sample matrix are often observed in the ICP–AES analysis of environmental matrices:

- High dissolved solids content in water (for example, chlorides and other salts in excess of 1500 mg/l) change the viscosity and the surface tension of the digested samples and affect the nebulization and aerosol transport.
- Another problem that can occur in samples with high dissolved solid contents is the salt buildup at the tip of the nebulizer that affects aerosol flow rate and causes instrumental drift.
- Because memory effects (carryover) from high concentration samples may take place in the nebulizer and in the plasma torch, the system must be flushed between samples. To recognize and prevent memory effects, instrument blanks are analyzed after high concentration samples; additional frequent instrument blanks intersperse analytical sequences as a QC measure.
- Plasma-induced high-temperature chemical reactions that may take place in the sample matrix have a potential to affect results. Laboratories must determine the presence of chemical interferences in unknown or chemically complex matrices and take measures to reduce them.

To detect and quantify the effects of physical and chemical interferences, the laboratories use a variety of techniques, such as *standard addition*, *serial dilution* or *internal standard addition*, which are described in Chapter 4.4.4.5.

4.4.4.4 The subtleties of digestion procedures

There are several sample digestion procedures used in elemental analysis. All of them use strong oxidizers (nitric acid, hydrochloric acid, and hydrogen peroxide) to solubilize environmentally available metals. The following distinctions between different types of elemental analysis digestion procedures are important for the planning of data collection and in data interpretation.

Total recoverable metals in aqueous samples

Total recoverable metals are defined as the concentrations of metals in an unfiltered aqueous sample treated with hot nitric and hydrochloric acids according to EPA Method 3005. The method is suitable for the preparation of digestates for the FLAA and ICP–AES analysis. This procedure is known to produce low recoveries of silver due to precipitation and of antimony due to evaporation.

Dissolved metals in aqueous samples

Dissolved metals are the concentrations of metals determined in a water sample after it has been filtered through a 0.45 μm filter and acidified with nitric acid immediately after filtering. Samples prepared in this manner do not need to be digested as long as the acidity of the sample has been adjusted to match the acidity of the standards. *Suspended metals* are defined as the concentrations of metals retained by the 0.45 μm filter. A sum of dissolved and suspended metal concentrations is the *total metals*.

Total metals in aqueous samples

Total metals are the concentrations of metals determined in a sample prepared according to EPA Methods 3010, 3015, 3020, 3050 or 3051. In EPA Method 3010, aqueous samples are digested with hot nitric and hydrochloric acids for FLAA and ICP–AES analyses. EPA Method 3020 is used for the preparation of aqueous samples for GFAA analysis. It is a less rigorous digestion procedure with hot nitric acid only. Excepted from EPA Method 3020 are arsenic, selenium, and silver, which are digested according to specific procedures described in their respective GFAA methods of the 7000 series.

Total metals in solid samples

Solid samples are digested for total metal analysis with hot nitric acid and hydrogen peroxide according to EPA Method 3050. The addition of hydrochloric acid is optional. There are two different digestion protocols within this method, one for FLAA and ICP–AES analyses and the other for GFAA analysis. These two procedures are not interchangeable, and for this reason samples prepared for FLAA or ICP-AES analysis must not be analyzed with GFAA methods.

Microwave digestion

Microwave-assisted digestion procedures are used for total metal analysis in aqueous samples (EPA Method 3015) and for solid or oily samples (EPA Method 3051). These procedures allow for a rapid sample digestion with nitric acid under high pressure and temperature conditions; the addition of hydrochloric acid is optional. Samples digested with microwave-assisted methods can be analyzed with FLAA, ICP–AES, GFAA or ICP–MS methods.

4.4.4.5   Matrix interference detection techniques

Several matrix interference detection techniques enable laboratories to recognize their presence and reduce their negative effects on compound identification and quantitation.

Serial dilution test

Laboratories use this test in ICP–AES and AA analysis of new and unknown or unusual matrices to determine if non-linear physical and chemical interferences may be obscuring the target analytes.

In this test, a digested sample is analyzed, and the digestate is then diluted by a factor of five and reanalyzed. Results of two analyses are compared, and if they are in agreement, the absence of interference is established. Different regulatory methods have different acceptance criteria for verifying the agreement between results. For drinking water analysis, the criteria are less than 10 percent for AA and less than 5 percent for ICP–AES methods. For hazardous waste analysis, a 10 percent agreement between results is acceptable for either AA or ICP–AES methods. This test is possible for the samples with the target analyte concentrations that are sufficiently high, at least 10–25 times of the IDL after the dilution.

If the difference between the two results exceeds the method criterion, physical and chemical interferences may be present, and laboratories should perform a post digestion spike addition to verify their presence.

Post digestion spike addition or recovery test

In this test, a known amount of a target analyte is added to a digested sample. The amount is between 10 and 100 times greater than the IDL. The sample is analyzed, and the recovery of the added analyte is calculated. If the recovery meets the method-specified acceptance criterion, the magnitude of matrix effects is defined, and results of analysis are acceptable. Similar to the serial dilution test, different regulatory methods have different acceptance criteria for verifying the agreement between results, such as the following:

- Drinking water analysis: 90–110 percent for AA and ICP–AES methods
- Hazardous waste analysis: 85–115 percent for AA methods and 75–125 percent for ICP–AES methods

If the recovery does not meet the specified criterion, the MSA is performed on all samples of the same matrix.

Method of standard additions

The MSA enables laboratories to calculate the analyte concentration in an interference-ridden sample. In the MSA technique, equal volumes of a digested sample are added to a blank and three standards that contain the target analyte at three different concentrations. The four preparations are analyzed, and the instrument response is plotted as the ordinate against the concentrations plotted on the abscissa. A straight line is drawn through the four points and is extrapolated to zero response on the abscissa. The intercept is the concentration of the analyte in the sample. Instead of plotting by hand, linear regression analysis is typically used to calculate the target analyte concentration. The addition of a single standard is also acceptable; the calculation of the target analyte concentration in this case is done based on two points instead of four.

Internal standards

An alternative to MSA in ICP-MS analysis is the internal standard technique. One or more elements not present in the samples and verified not to cause an interelement spectral interference are added to the digested samples, standards, and blanks. Yttrium, scandium, and other rarely occurring elements or isotopes are used for this purpose. Their response serves as an internal standard for correcting the target analyte response in the calibration standards and for target analyte quantitation in the samples. This technique is very useful in overcoming matrix interferences, especially in high solids matrices.

Practical Tips: Methods for elemental analysis

1. Remember that elemental analysis methods cannot distinguish between the oxidation states of the elements, for example, between Se^{4+} and Se^{6+} or between As^{3+} and As^{5+}.

2. When visiting a laboratory, note if elemental analysis areas are free of dust and dirt from handling soil samples. A dirty trace element analysis laboratory is a bad laboratory.

3. When results of ICP–AES analysis indicate possible interelement interference, request confirmation with AA methods.

4. To reach the low detection limits and eliminate the possibility of matrix interferences, choose the GFAA methods for analysis of antimony, arsenic, beryllium, cadmium, chromium, lead, selenium, silver, and thallium.

5. Although ICP–MS is capable of analyzing over 60 elements, the EPA has approved only 15 elements for inclusion in EPA Method 6020.

6. Be specific in your instructions to the laboratory regarding digestion and analysis methods: subtle differences in procedures may produce dramatic differences in results. For example, antimony results are greatly affected by the digestion method, as it easily volatilizes at elevated temperatures and forms insoluble oxides in a reaction with hydrogen peroxide.

7. For better navigation through the metal analysis methods of SW-846, remember that FLAA methods are usually numbered with even numbers and GFAA methods with odd ones.

4.5   COMPOUND QUANTITATION

Chapter 4.4 addressed various aspects of qualitative identification of pollutants with common analytical techniques, such as chromatography and elemental analysis. Another integral component of environmental analysis is pollutant quantitation. For the data to be valid and usable, the analytes must be not only correctly identified but also properly quantified.

Two important components of quantitative analysis of environmental samples are the *determination of method detection limits* and *instrument calibration*. Understanding how they contribute to data quality will enable us to make decisions related to data validity during the assessment phase of the data collection process.

4.5.1   Detection and quantitation limits

Existing definitions of various detection and quantitation limits can be confusing to a non-laboratory person. Despite misleading similarities of these definitions, there is a logic and order to the basic concepts that they express. Various detection limits that we commonly refer to in our daily work (the IDLs, MDLs, and PQLs) are discussed in this chapter in the increasing order of magnitude of their numeric values. Some of these detection limits are determined experimentally and depend on the matrix and the method of preparation and analysis, while others may be arbitrary values selected by the laboratory or the data user. The relationship between these three levels of detection is approximately 1:5:10.

Instrument detection limit

Environmental laboratories routinely determine IDLs in the course of ICP–AES analysis as a measure of background and interelement interferences at the lowest measurable concentration level above the background noise. *The IDL is a trace element analyte concentration that produces a signal greater than three standard deviations of the mean noise level or that can be determined by injecting a standard to produce a signal that is five times the signal to noise ratio* (APHA, 1998).

Consistent with these definitions, there are two methods for IDL determination. The first method consists of multiple analyses of a *reagent blank*, followed by the determination of the standard deviation of the responses at the wavelength of the target analyte. The standard deviation multiplied by a factor of three is the IDL. This calculation defines the IDL as an analyte signal that is statistically greater than the noise.

The laboratories, however, typically rely on the second method for the IDL determination, which is detailed in the CLP SOW (EPA, 1995c). The second method consists of multiple analyses of a *standard solution* at a concentration that produces a signal five times over the signal-to-noise level. The standard deviation of the measurements is multiplied by a factor of three to produce the IDL. This method assumes that the level of signal-to-noise is known, and this information is usually available from the instrument manufacturer.

In ICP–AES analysis, the IDLs must be determined every time an instrument has been adjusted in a way that may affect its sensitivity and interelement correction. The CLP SOW requires that the IDLs be determined at a minimum on a quarterly basis. Laboratories that are not part of the CLP determine IDLs annually. Table 4.4 shows a comparison of the IDLs for selected elements obtained with various elemental analysis techniques.

Method detection limit

The MDL is one of the secondary data quality indicators. The EPA provides the definition of the MDL as *the minimum concentration that can be measured and reported with 99 percent confidence that the analyte concentration is greater than zero* (EPA, 1984a).

To determine the MDLs, laboratories prepare and analyze seven replicates of an analyte-free matrix (reagent water and laboratory-grade sand) spiked with the target analytes at concentrations that are 3–5 times greater than the estimated MDLs. Using seven results, the laboratories calculate the MDL according to the following equation:

$$MDL = t \times s$$

where t is the Student's t value of 3.143 for six degrees of freedom and a 99 percent confidence interval and s is the standard deviation of the seven results.

More than seven replicates may be analyzed; however, all of the obtained results must be used in the calculation, unless there is a well-justified reason to discard any of them. MDLs are specific to a given matrix, method, and instrument, and greatly depend on the analyst's technique. The better the analytical precision, the lower the calculated value of the MDL. Laboratories are required to perform MDL studies at least once a year (APHA, 1998; EPA, 1999d). However, the MDLs may be determined more often if there is a change in the laboratory extraction, analysis, or instrumentation. For trace element analyses, the MDL studies are performed in reagent water only, as a metal-free solid matrix that would successfully emulate natural soils does not exist.

Estimated quantitation limit or practical quantitation limit

The terms *estimated quantitation limit (EQL)* and *practical quantitation limit* describe the limit of quantitation, another secondary data quality indicator. These terms are used interchangeably. In fact, the common term used by the laboratories is the *PQL*. The EPA, however, prefers to use the term *EQL* and defines it as follows: *The estimated quantitation limit (EQL) is the lowest concentration that can be reliably achieved within specified limits of precision and accuracy during routine laboratory operating conditions* (EPA, 1996a). The PQL is defined similarly (EPA, 1985).

Typically, laboratories choose the analyte PQL value at 2–10 times its MDL. The selection, however, is not entirely arbitrary because the laboratories must use the selected PQL concentration value as the lowest standard in the multipoint calibration curve. This enables the laboratory to assure that even at a low concentration level, the analyte is detected, identified, and quantified correctly. Therefore, the PQL may be also defined as *a concentration that is 2–10 times greater than the MDL and that represents the lowest point on the calibration curve during routine laboratory operations*.

Once the PQLs have been established, laboratories use them as routine reporting limits in the analysis of interference-free, undiluted samples. The PQLs, however, are highly matrix-dependent and their values increase with sample dilution.

Reporting limit

Reporting limit is a PQL for a specific analyte; it is calculated as the PQL multiplied by the dilution factor. The RL may be equal to the PQL or greater than the PQL. An example of a PQL transformation into an RL is shown in Table 4.2.

As an exception, the RL may be equal to the MDL. This may happen when very low RLs would be appropriate for the intended use of the data, but current analytical methods do not have sufficient sensitivity to achieve them. In this case, it is not unusual to have the data reported using the MDLs as the RLs. This creates a unique situation when the RL value is below the PQL value. Because of the analytical uncertainty associated with the concentration values between the MDL and the PQL, these concentrations are regarded as estimated values.

Contract required quantitation limits

The CRQLs are the reporting limits specified in the CLP SOW for organic and inorganic analyses. The EPA selected these reasonable reporting limits based on the capabilities of current analytical instruments and standard laboratory practices.

The concept of the CRQL extents beyond the CLP SOW. The SAP and the Laboratory SOW described in Chapter 2.8 should list the reporting limits required for the project DQOs. Once a contract with the laboratory has been signed, these reporting limits become the *de facto* CRQLs. These limits, however, should not be chosen at will and there must be a scientific basis for their selection. The published analytical methods usually include the EQLs or PQLs, which we can use as the baseline information of a method's detection capability. Laboratories have their own MDLs and PQLs, which are specific to the analytical procedures and instruments. As a rule, laboratories provide the MDL and PQL information to their existing and prospective clients to assist them in the development of the SAP.

Practical Tips: Detection and quantitation limits

1. Select the PQLs based in the project DQOs with the intended use of the data guiding the selection.

2. Specify the PQLs appropriate for your project work in the planning documents and in the Laboratory SOW.

3. Do not request the lowest achievable PQLs unless they are justified by the DQOs as this unnecessary requirement may increase the price of analysis.

4. Compare the project action levels to the PQLs: the PQLs should be at least 2 times lower than the corresponding action levels.

5. Prepare a table for the SAP that compares the action levels to the PQLs.

Practical Tips: Detection and quantitation limits (continued)

6. If the PQLs of routine analytical methods are still above the action levels, contact the laboratory to discuss the possibility of developing project-specific analytical procedures with lower PQLs.

7. If available analytical methods are not sufficiently sensitive to meet the project DQOs, request that the laboratory use the MDLs as reporting limits. Concentrations between the MDL and the PQL will be estimated values.

8. Ask the laboratory for the MDL study documentation; a laboratory without documented MDL studies cannot be trusted to produce valid data.

9. Some effects of matrix interferences that cause elevated PQLs cannot be overcome even with the leading edge preparation and analysis techniques.

4.5.2   Importance of calibration

Before analysis may start, the instrument must be calibrated. *Calibration is the process of standardizing an instrument's response in order to perform quantitative analysis.* Laboratories must perform initial calibration prior to the start of sample analysis and confirm its validity throughout sample analysis with calibration verification checks.

4.6.2.1   Rules of calibration

Initial calibration is a multipoint correlation between several analyte concentrations in standard solutions and the instrument's response. This multipoint correlation, referred to as the *calibration curve*, may be mathematically expressed as a linear or non-linear regression. ICP–AES instruments are excluded from this requirement since they have the software that does not allow multipoint calibrations to be performed.

In organic compound analysis, the instrument response is expressed as a *response factor* (RF), which is the ratio of the concentration (or the mass) of the analyte in a standard to the area of the chromatographic peak. Conversely, a *calibration factor* (CF) is the ratio of the peak area to the concentration (or the mass) of the analyte. Equation 1, Appendix 22, shows the calculation of RF and CF. In trace element and inorganic compound analyses, the calibration curve is usually defined with a linear regression equation, and response (calibration) factors are not used for quantitation.

Different analytical techniques and methods have different calibration requirements, but there are common rules to all of them.

Rule 1. All target analytes must have acceptable calibration curves.

• Linear calibrations similar to one shown in Figure 4.6 are prevalent in all types of analysis.
• For linear calibrations, compounds may be quantified using the average response factor or the average calibration factor (Equation 2, Appendix 22). The measure of linearity is the RSD, calculated according to Equation 3, Appendix 22.

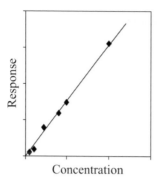
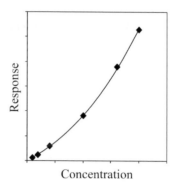

Linear calibration curve with an RSD   Calibration curve with a quadratic fit and
of less than 20 percent   a coefficient of determination of 0.999

Figure 4.6   Linear and non-linear multipoint calibration curves.

- The RSD for *each target analyte* in the calibration curve must meet the method acceptance criteria before analysis may start.
- Acceptable calibrations for indicator compounds, such as the **calibration check compounds** (CCCs) in EPA Methods 8260 and 8270, shown in Example 4.14, do not eliminate the need for acceptable calibrations of the target analytes.
- Linear calibration models may be based on a linear regression with the correlation coefficient that is greater than or equal to 0.99.
- The non-linear calibration models are also acceptable. Figure 4.6 shows an example of a non-linear calibration approximated with a quadratic fit. Non-linear calibration curves are not acceptable if used to compensate for the detector saturation at a high concentration level.
- Non-linear polynomial calibration models must have the coefficient of determination greater than or equal to 0.99.
- A linear (first order) calibration model requires five standards, a quadratic (second order) model requires six standards, and a third order polynomial calibration model requires seven standards.
- Laboratories calibrate ICP–AES instruments according to manufacturer's instructions by analyzing a calibration blank and one standard to span the full dynamic linear range of the instrument.
- Similar to target analytes, surrogate standards must be calibrated with multipoint calibration curves.
- GC/MS instruments must have an acceptable tuning check with BFB or DFTPP before they are calibrated.

Rule 2. After a calibration curve has been created, a single point initial calibration verification with a standard of a different source must take place.

- This is a critical but often overlooked rule of calibration. A systematic error in the calibration standard preparation is likely to remain undetected unless a standard of

---

Example 4.14: **Calibration check compounds for EPA Methods 8260B and 8270C**

The purpose of the CCCs in GC/MS analysis is to evaluate the calibration from the standpoint of the integrity of the analytical system. The EPA Method 8270 analytes include base, neutral, and acid compounds (BNAs), which have different chemical behavior, and the CCCs include the representatives of all three groups.

CCCs—EPA Method 8260B

- 1,1-Dichloroethene
- Chloroform
- 1,2-Dichloropropane
- Toluene
- Ethylbenzene
- Vinyl chloride

CCCs—EPA Method 8270C

Base/Neutral Fraction
- Acenaphthene
- 1,4-Dichlorobenzene
- Diphenylamine
- Hexachlorobutadiene
- Di-*n*-octyl phthalate
- Fluoranthene
- Benzo(a)pyrene

Acid Fraction
- 4-Chloro-3-methylphenol
- 2,4-Dichlorophenol
- 2-Nitrophenol
- Phenol
- Pentachlorophenol
- 2,4,6-Trichlorophenol

High response variability for these compounds may indicate system leaks or the presence of reactive sites in the column. According to EPA Methods 8260B and 8270C, the RSD for each individual CCC must be less than 30 percent, whereas the RSD for target analytes must be less than 15 percent. This means that the CCCs, even if they are target analytes, may have wider calibration acceptance criteria than the rest of the target analytes. If an RSD exceeding 30 percent has been determined for any CCCs, corrective action to eliminate a system leak and/or column reactive sites is necessary before reattempting the instrument calibration.

---

a different source is used as a QC check. The practice of ICV is necessary for determining errors that originate from incorrect analyte concentrations in the initial calibration standards and for assuring correct quantitation.

- At least one of the standards (the one used for initial calibration or the ICV standard) must be certified traceable to a national standard, such as the NIST-traceable standard reference materials. Certified stock standard solutions for the majority of environmental pollutants are available commercially. However, whether a stock standard is a certified solution or has been prepared by the laboratory, dilution errors during the standard preparation and analyte degradation in storage happen. The laboratories are able to detect such errors *only* by preparing and analyzing an independent, second source confirmation standard.
- The ICV check standard has the acceptance criteria that must be met before the analysis may start.
- A linear calibration ICV acceptance criterion is ***percent difference (%D)***, calculated as shown in Equation 4, Appendix 22.
- A linear regression or non-linear calibration ICV acceptance criterion is ***percent drift*** calculated as shown in Equation 5, Appendix 22.

- Linear calibration verification is always performed with a single ICV standard. Non-linear calibrations may require two ICV standards analyzed at different inflections of the curve.

Rule 3. Continuing calibration verification must take place at regular intervals during sample analysis.

- A continuing calibration verification standard at a concentration that is close to the midrange of the calibration curve may be prepared either from the same stock solution as the initial calibration standard or from a standard of a different source. This CCV standard is analyzed at a certain frequency throughout sample analysis.
- The acceptability of the CCV is determined based on the percent difference or percent drift calculated according to Equations 4 and 5, Appendix 22.
- The CCV response (calibration) factor has the acceptance criteria that must be met for *each target analyte* before the analysis may proceed.
- The CCV response (calibration) factor cannot be used for compound quantitation, only the average response (calibration) factor of the initial calibration can. The only exceptions to this rule are the CLP SOW methods for organic compound analyses that allow use of CCV for compound quantitation.
- For organic compound analyses, the accepted practice based on SW-846 requirement is to verify the initial calibration at the beginning of every 12-hour shift. For trace element and inorganic compound analyses, a CCV must be analyzed after every tenth sample and at the end of the analytical sequence.
- A series of analyzed samples must be bracketed by acceptable CCVs for the analysis to be valid. If any of the CCVs for the target analytes do not meet the acceptance criteria, the laboratory must take corrective action, reconfirm the initial calibration validity, and reanalyze the samples.
- Linear calibration verification is always performed with a single CCV standard. Non-linear calibrations may require two CCV standards analyzed at different inflections of the curve.
- For analysis to begin, GC/MS instruments must have acceptable tuning checks with BFB or DFTPP analyzed at a method-specified frequency.

Rule 4. The lowest calibration standard analyzed during initial calibration establishes the method PQL based on the final volume of extract or sample.
This rule is best explained using real data, shown in Example 4.15.

Rule 5. To produce valid data, the response of the analyte in the sample must fall within the calibrated range of instrument responses.

- Extrapolation of the calibration curve above or below the calibrated range is not acceptable. The rule of thumb is that samples with analyte concentrations that exceed the upper calibration point by more than 10 percent must be diluted. Sample concentrations that are below the lower calibration point are estimated, not definitive, values.

Example 4.15: PQL and the lower calibration point

- In the purge and trap analysis, if the PQL of an analyte is $2\,\mu g/l$ based on a 5 ml sample volume, then the lowest standard in the calibration curve must also be at a concentration of $2\,\mu g/l$ and have a volume of 5 ml.
- Suppose that a lower PQL is required for a project, and the laboratory chooses to use a greater sample volume (for example, 25 ml), assuming a five-fold improvement in the PQL ($0.4\,\mu g/l$). In this case, a new calibration curve must be created using a 25 ml purge volume of all standard solutions with the lowest concentration point at $0.4\,\mu g/l$.
- For a method that requires sample extraction, consider pesticide analysis with EPA Method 8081. If the PQL for an analyte is $0.05\,\mu g/l$, then the lower calibration level must be $0.05\,\mu g/ml$, provided that a sample volume of one liter is extracted and reduced to 1 ml. For soil analysis, the PQL for this analyte will be $17\,\mu g/kg$, provided that the weight of extracted soil is 30 g and the final volume of the extract is 10 ml.

- The laboratories maximize their sample throughput by reducing the number of samples that need reanalysis due to dilution. To reach this end, they routinely calibrate all of the instruments using the widest possible calibration range that spans the full linear dynamic range of the detector.
- On the other hand, a wide calibration range that extends into the parts per million or even percent range is not appropriate for the analysis of samples with the target analytes in the parts per billion range. A different calibration curve must be created for such samples.

Rule 6. Different regulatory protocols have different calibration requirements.

- Different analytical methods used for the analysis of samples collected under the requirements of different environmental laws are discussed in Chapter 2.4. Although many of these methods target the same analytes, their calibration requirements are different. Tables 4.5, 4.6, and 4.7 summarize the differences in calibration requirements for organic compound and trace element analysis. (Inorganic analyte methods and techniques have a range of requirements that cannot be summarized in a concise manner; we should refer to specific methods for this information).
- Regulatory method requirements should be examined and understood by the laboratory and the data user during the planning phase of the project. These requirements must be summarized in the SAP, as they become the criteria for establishing data validity and usability in the assessment phase of the project.
- Environmental laboratories have their procedures established in a manner that allows the most efficient sample throughput. This means that the requirements of various regulatory methods may be converged into a single laboratory protocol that meets the most stringent of all requirements. Specific method requirements that are not addressed in these uniform protocols, must be, nevertheless, met during analysis of samples collected to satisfy the provisions of a given environmental law.

Table 4.5 Calibration requirements for GC and HPLC methods

Initial calibration	Initial calibration verification with a second source standard	Continuing calibration verification
Wastewater, EPA 600 Series (EPA, 1983)		
Three-point with the RSD<10%	Difference ±15%	• Daily • Difference ±15%
Drinking water, EPA 500 Series (EPA, 1992)		
Three to five-point with the RSD<10% or 20%, depending on the method	Difference ±20%	• Daily • Difference ±20%
Solid waste, EPA 8000 Series (EPA, 1996a)		
• Five-point with the RSD ≤20% • Correlation (determination) coefficient ≤0.99	Difference or drift ±15%	• Every 12 hours • Difference or drift ±15%

Table 4.6 Calibration requirements for GC/MS methods

Initial calibration	Initial calibration verification with a second source standard	Continuing calibration verification
Wastewater, EPA 600 Series (EPA, 1983)		
Three-point with the RSD<35%	Acceptability limits specified for each analyte separately	• Daily tuning check with BFB or DFTPP • Daily calibration check with acceptability limits specified for each analyte separately
Drinking water, EPA 500 Series (EPA, 1992)		
Three to five-point with the RSD<20%	Quarterly; acceptability limits not specified	• Tuning check with BFB or DFTPP every 8 hours • Calibration check every 8 hours; difference ±30%
Solid waste, EPA 8000 Series (EPA, 1996a)		
• Five-point with the RSD≤15% for target analytes • RSD≤30% for CCCs	Difference monitored only for the CCCs and is ±20%	• Tuning check with BFB or DFTPP every 12 hours • Calibration check every 12 hours • Difference monitored only for the CCCs and is ±20%

Table 4.7 Calibration requirements for trace element analysis methods

Initial calibration	Initial calibration verification with a second source standard	Continuing calibration verification
Wastewater and drinking water, EPA 600 Series (EPA, 1983)		
EPA Method 200.7 (ICP–AES): one standard and a blank	Drift ±5%	• Every 10 samples • Drift ±5%
EPA 200 Series (AA): three standards and a blank with a correlation coefficient ≤0.995	One standard near the MCL and a blank Drift ±10%	• One standard near the MCL and a blank every 20 samples • Drift ±10%
Solid waste, EPA 8000 Series (EPA, 1996a)		
EPA Method 6010 (ICP–AES): one standard and a blank	Drift ±10%	• Every 10 samples • Drift ±10%
EPA 7000 Series (AA): three standards and a blank with a correlation coefficient ≤0.995	One midrange standard and a blank Drift ±10%	• One midrange standard and a blank every 10 samples • Drift ±20%

4.5.2.2 Internal and external standard calibration

Analytical instruments may be calibrated using either external or internal standard technique.

External standard calibration

External standard calibration is used if the changes in the analytical system that may occur from the time the instrument has been calibrated to the time the sample analysis is completed are negligible. These changes are assumed to produce an insignificant error that is built into the daily calibration verification acceptance criteria. The response (calibration) factor of the external standard calibration is calculated according to Equation 1, Appendix 22, and it is measured in concentration or mass units (or their inverse).

Internal standard calibration

Internal standard calibration is used when the changes in the analytical system are known to be frequent and substantial. To compensate for these changes, *internal standards* at known concentrations are added to all calibration standards, field samples, and laboratory QC samples prior to analysis. ***Internal standards are synthetic analogs of specific target compounds or compounds that are similar in nature to the target analytes and that are not found in environmental samples.*** Internal standard calibration is a requirement of GC/MS methods. Laboratories sometimes use it for GC methods as it significantly improves the accuracy of compound quantitation.

During the preparation of standards for internal standard calibration, the same volume and concentration of the internal standard is added to every calibration standard. The concentration of the internal standards stays constant, whereas the concentrations of the calibration standards vary. The same volume and concentration of the internal standard is added to every field sample and laboratory QC sample prior to purge and trap analysis or to sample extracts for direct injection analysis. The volume of the internal standard added must be negligible compared to the volume of the calibration standard or the sample.

The response factor for each analyte in the calibration standard, often referred to as the *relative response factor*, is calculated relative to the response of the internal standard as shown in Equation 9, Appendix 22. Unlike the response (calibration) factor used in the external standard calculations, relative response factor is unitless. Once the relative response factors have been calculated, the average relative response factor, the RSD, the percent difference or drift are calculated according to Equations 2–5, Appendix 22.

Analytical system performance monitoring based on the response factor of several representative compounds is an excellent diagnostic QC tool. This type of monitoring is a requirement in GC/MS methods, as illustrated in Example 4.16, and it may be applied to other analytical methods.

Example 4.16:   System performance check compounds in EPA Methods 8260 and 8270

During VOC and SVOC analysis with GC/MS, the magnitude of the relative response factor is monitored throughout the analysis. This is a quality check that allows the analyst to identify compound degradation caused by active sites in the system and by compound instability. Following are system performance check compounds (SPCCs) and their minimum RFs for EPA Methods 8260 and 8270 (EPA, 1996a):

EPA Method 8260	Minimum RF	EPA Method 8270	Minimum RF
Chloromethane	0.10	n-Nitroso-di-n-propylamine	0.05
1,1-Dichloroethane	0.10	Hexachlorocyclopentadiene	0.05
Bromoform	0.10	2,4-Dinitrophenol	0.05
Chlorobenzene	0.30	4-Nitrophenol	0.05
1,1,2,2-Tetrachloroethane	0.30		

4.5.2.3   Calculations of sample concentrations

To perform an external or internal standard calibration procedure and to calculate sample results, the laboratory conducts a series of calculations. Although these calculations are computerized at a modern laboratory, professional judgment and experience of the analyst are critical in the selection of the appropriate calibration model and for correct programming of the data acquisition system that collects the data and performs the calculations.

Understanding how the calculations are conducted is important for the assessment of data quality, as the recalculation of results from raw data may disclose an undetected laboratory error. In this chapter, we will review common internal and external standard calculations.

The equations for compound quantitation are shown in Appendix 22. Most of these equations apply to linear calibration models that rely on average response (calibration) factor for compound quantitation. Calibrations that use linear regression and non-linear polynomial equations read compound concentrations in the analyzed sample aliquot directly from the calibration curve. Once this concentration has been obtained, the final sample concentration can be calculated using the same rationale as for the linear concentration model.

There are several ways to calculate compound concentrations from a calibration curve, and Appendix 22 shows examples of how these calculations may be conducted. Laboratories may use different equations, but whichever equation is used, the final result should be always the same.

Linear external or internal calibration analytical sequence

1. Analyze calibration standards
2. Calculate the instrument RF or CF: Equation 1 or 9, Appendix 22
3. Calculate the average RF or CF: Equation 2, Appendix 22
4. Calculate the RSD: Equation 3, Appendix 22
5. Compare the RSD to initial calibration acceptance criteria: Tables 4.5, 4.6, and 4.7
6. Perform the ICV, calculate percent difference: Equation 4, Appendix 22
7. Compare percent difference to acceptance criteria: Tables 4.5, 4.6, and 4.7
8. Analyze samples, calculate results:

   - For direct injections of sample extracts: Equation 6 or 10, Appendix 22
   - Purge and trap analysis of water and low level soil samples: Equation 7 or 11, Appendix 22
   - Purge and trap analysis of medium and high level samples that require a solvent extraction before analysis: Equation 8 or 12, Appendix 22

9. Perform the CCV, calculate percent difference: Equation 4, Appendix 22
10. Compare percent difference to acceptance criteria: Tables 4.5, 4.6, and 4.7

Linear regression and non-linear calibration analytical sequence

1. Analyze calibration standards
2. Find the best fit that meets the initial calibration acceptance criterion (Tables 4.5, 4.6, and 4.7); program the data acquisition computer to use it for sample calculation
3. Perform the ICV, calculate percent drift: Equation 5, Appendix 22
4. Compare percent drift to acceptance criteria: Tables 4.5, 4.6, and 4.7
5. Analyze samples, calculate results: Equation 13, Appendix 22
6. Perform the CCV, calculate percent drift: Equation 5, Appendix 22
7. Compare percent drift to acceptance criteria: Tables 4.5, 4.6, and 4.7

Practical Tips: Importance of calibration

1. Remember the basic rules: organic methods typically require a five-point calibration; trace element AA methods use a calibration blank and three standards; the ICP–AES instruments are calibrated with one standard and a calibration blank; inorganic analysis methods use three- to five-point calibration curves.

2. Chromatography methods employ multipoint (5–7 points) calibration curves.

3. Specify calibration requirements in the SAP and in the Laboratory SOW by referencing applicable environmental regulations and analytical methods.

4. Some instruments maintain initial multipoint calibration curves current longer than others. For example, for organic methods, an initial calibration may serve from one month to half a year. However, even under the most careful laboratory conditions, sooner or later instruments will require a new multipoint calibration.

5. When in doubt about the data quality, request that the laboratory provide calibration data for the reconstruction of result calculations.

6. The date of initial calibration should be clearly stated in the data package. If the initial calibration cannot be traced, the data cannot be considered valid.

7. The laboratory that cannot reproduce its own calculations or is unable to supply the calibration information to the data user cannot be trusted to produce valid data.

8. Because numerical values for a CCV (ICV) percent drift and percent difference of a linear calibration are almost the same, laboratories often use percent drift as the acceptance criteria for linear calibrations.

9. Mass spectrometer tuning acceptance criteria are different for different regulatory methods.

4.6   LABORATORY QUALITY ASSURANCE

Quality assurance is a set of operating principles that enable laboratories to produce defensible data of known accuracy and precision. These operating principles that form laboratory quality system are documented in the Laboratory QA Manual, in a set of laboratory SOPs, and in various laboratory records.

Two elements of quality assurance are *quality control* and *quality assessment*. *Quality control is a set of measures implemented within an analytical procedure to assure that the process is in control.* A combination of these measures constitutes the laboratory QC program. A properly designed and executed QC program will result in a measurement system operating in a state of statistical control, which means that errors have been reduced to acceptable levels. An effective QC program includes the following elements:

- Initial demonstration of capability
- MDL determinations
- Calibration with standards that are traceable to NIST-certified reference materials

- Definition of a laboratory batch
- Analysis of laboratory blanks
- Analysis of surrogate standards and laboratory control samples to measure analytical accuracy
- Analysis of laboratory duplicates to measure analytical precision
- Maintenance of control charts to determine the acceptance criteria for accuracy and precision
- Corrective action to resolve the non-conforming events

Quality assessment is the process used by the laboratory to ensure that the QC measures are implemented correctly and that the quality of produced data is determined. Included in quality assessment are internal and external audits of laboratory systems and analysis of performance evaluation samples.

Every member of laboratory operations is responsible for the implementation of the QC program, whereas QA oversight is conducted by the QA section staff and is independent of laboratory operations.

A typical data user is only interested in those QC checks that determine the quality of a particular set of data. Internal laboratory processes, such as initial demonstration of capability or corrective action, are not a concern of a data user unless a gross analytical error has been discovered that casts doubt over the effectiveness of the laboratory quality system.

4.6.1 Laboratory quality control samples and their meaning

The laboratory uses QC check samples for the calculations of the primary DQIs (accuracy and precision). In addition, QC check samples enable laboratories to monitor the secondary DQIs, such as memory effects and recovery. The purpose and frequency of analysis of the laboratory QC check samples are summarized in Table 4.8.

Different regulatory methods have different requirements for the type and frequency of analysis of laboratory QC check samples. That is why it is important to specify these requirements in the SAP.

Laboratory QC data are classified as **batch QC data** and **individual sample QC data**. For all types of analysis, batch QC data originate from laboratory blanks, laboratory control samples, matrix spikes, and laboratory duplicates. Individual sample QC data in organic compound analysis are obtained from surrogate and internal standard recoveries. Matrix interference detection techniques (serial dilution tests, post-digestion spike additions, and MSA tests) are the source for individual sample QC checks in trace element analysis. (Chapter 4.4.4.5 addresses the trace element matrix interference detection techniques and the associated acceptance criteria.)

Included in the batch QC data are method-specific QC checks, such as GC/MS tuning checks and dichlorodiphenyltrichloroethane (DDT) and Endrin degradation checks listed in Table 4.8. These checks are necessary to evaluate the analytical system condition prior to and during analysis. GC/MS tuning checks with BFB and DFTPP solutions described in Section 4.4.1.1 are necessary for obtaining comparable data from all laboratories performing analysis with GC/MS methods. The DDT and Endrin degradation checks assure the accurate quantitation of these pesticides and their natural degradation products.

Table 4.8   The purpose of laboratory QC check samples

Purpose	Frequency of analysis
Calibration blank	
A point in a calibration curve	• Part of initial calibration in trace element and inorganic compound analysis
Instrument blank	
Determines whether memory effects (carryover) are present in the analytical system	• Every 10 samples in trace element analysis • After high concentration samples in organic analysis
Method blank	
Establishes that laboratory contamination does not cause false positive results	• Prepared with every preparation batch of up to 20 field samples for all organic, inorganic, and trace element analysis • Analyzed once as part of the analytical batch
Laboratory control sample/laboratory control sample duplicate	
Determine analytical accuracy and precision for a sample batch	• One pair for every preparation batch of up to 20 field samples
Matrix spike	
Determines the effects of matrix interferences on analytical accuracy of a sample	• Water: depends on the project DQO • Soil: performed at least once for every 20 field samples
Matrix spike duplicate	
Determines the effects of matrix interferences on analytical accuracy of a sample; together with matrix spike determines analytical precision	• Depends on the project DQOs
Laboratory duplicate	
Determines analytical precision of a sample batch	• One for every 10 samples for trace element and inorganic analysis
Surrogate standard	
Determines analytical accuracy for each sample	• Added to every blank, field, and laboratory QC sample
Internal standard	
Compensates for the changes in the analytical system in internal standard calibration procedure	• Added to every standard, blank, field, and laboratory QC sample
GC/MS tuning check	
Verifies the performance of the GC/MS system	• Before samples are analyzed • Every 12 hours thereafter or as required by the method (see Table 4.6)
DDT and Endrin breakdown product monitoring in organochlorine pesticide analysis	
Detects contamination, active sites, and unacceptably high temperatures within the analytical system during pesticide analysis by EPA Method 8081	• Before samples are analyzed • Every 12 hours thereafter

4.6.1.1 Laboratory batch
Before laboratory QC checks are discussed, we need to revisit the concept of batch analysis introduced in Chapter 4.3. Laboratories process samples in a batch manner by assembling sets of samples that are prepared and analyzed together. Batch analysis enables laboratories to maximize sample throughput while performing a minimum of the required laboratory QC checks.

- *A preparation batch is a group of up to 20 field samples, prepared together for the same analysis using the same lots of reagents and associated with common QC samples.* In addition to field samples, a preparation batch must, at a minimum, include a method (extraction or digestion) blank, an LCS, and an LCSD. Other laboratory QC checks may be part of the preparation batch, such as an MS/MSD pair or a laboratory duplicate. If laboratory QC checks in a preparation batch meet the laboratory acceptance criteria, the batch is considered be in a state of control and every sample in it is acceptable, provided that individual QC checks are also acceptable. If the method blank and the samples in a preparation batch show contamination that makes sample results inconclusive or if the LCS and LCSD recoveries are not acceptable, the whole batch may be prepared again.
- *An analytical batch is a group of samples, extracts or digestates, which are analyzed sequentially using the same calibration curve and which have common analytical QC checks.* These are the samples that are bracketed by the same CCVs, have the same instrument blanks, and other QC checks that may be required by the method (for example, the DDT and Endrin breakdown product check in organochlorine pesticide analysis by EPA Methods 8081). If the CCV or any of the analytical QC checks are outside the method acceptance criteria, the whole analytical batch or only the affected samples are reanalyzed.

The preparation or analytical batch should not be confused with the Sample Delivery Group or SDG, a CLP SOW term used to describe a group of up to 20 samples of the same matrix received at a CLP laboratory within a 14-day period. Non-CLP laboratories often use this term interchangeably with the Work Order or Laboratory Project Number that they assign to a group of samples from the same COC Form.

Preparation and analytical batches must be clearly identified with a unique number in laboratory bench sheets, notebooks, and computer systems. The same applies to QC check samples associated with each batch. During data quality assessment, the data user will determine the quality of the field sample data based in the results of the batch QC check samples that are part of the preparation and analytical batches. The data user will examine batch QC check samples first and, if they are acceptable, will proceed to individual sample QC checks. A complete examination of these QC checks will enable the data user to evaluate the quantitative DQIs (accuracy and precision). A combination of acceptable batch QC checks and individual QC checks makes the data valid on condition that the qualitative DQIs (representativeness and comparability) are also acceptable.

4.6.1.2   Laboratory blanks

Laboratory blanks are batch QC checks used to demonstrate that laboratory contamination or residual contamination in the analytical system (memory effects) does not cause false positive results. Laboratory blanks include *calibration*, *instrument*, and *method* blanks.

The calibration blank in trace element analysis is prepared by acidifying reagent water with the same concentrations of the acids used in the preparation of standards and samples. It serves as a calibration point in the initial calibration. As part of an analytical batch, the calibration blank is analyzed frequently to flush the analytical system between standards and samples in order to eradicate memory effects. Calibration blanks are also used in inorganic compound analysis, where they are prepared with the chemicals specified by the method.

The instrument blank is the reagent used in the final preparation of the samples, which is injected into the analytical system after high level standards and samples have been analyzed. Analysis of instrument blanks removes memory effects from the analytical system. A calibration blank used for flushing the system in trace element analysis is an example of an instrument blank.

The method blank is the analyte-free matrix (reagent water or laboratory-grade sand) that is carried throughout the entire preparation and analysis and contains all of the reagents in the same volumes as used in the processing of the samples. A method blank is part of every preparation batch of up to 20 field samples. A method blank is analyzed after a clean instrument blank to confirm that laboratory contamination does not cause false positive results and to demonstrate that contamination, if present, is not a result of memory effects. Method blank contamination can be a serious breach of the QC protocol, often warranting a new preparation of the whole batch. *Subtracting method blank results from sample results is an unacceptable practice.* Generally, a method blank is considered acceptable if no target analytes are detected above the PQL.

4.6.1.3   Laboratory control samples

Laboratory control samples are analyte-free matrices (reagent water or laboratory-grade sand) fortified (spiked) with known concentrations of target analytes and carried throughout the entire preparation and analysis. Laboratories prepare and analyze these batch QC check samples at minimum frequency of one LCS/LCSD pair for every preparation batch of up to 20 field samples. Laboratory control samples serve two purposes:

1. To evaluate analyte recovery from an interference-free matrix as a measure of analytical accuracy
2. To determine laboratory precision as the RPD between the LCS and the LCSD recoveries

   A laboratory that does not prepare and analyze LCS/LCSD routinely does not have the means to evaluate analytical accuracy and precision. Consequently, the quality of data produced at such a laboratory cannot be determined.

The spiking solution for the LCS/LCSD preparation should preferably contain compounds that are expected to be present in the samples, however, these are hard to predict. The EPA-recommended spiking solutions for several methods, shown in Example 4.17, include a minimum number of analytes (EPA, 1996a). Laboratories either follow these recommendations or use a different mixture of spiking compounds that may even include a full list of a method's target analytes. Most importantly, the spiking solution should represent the different chemical classes that are on the target analyte list. For example, EPA Method 8270 analytes include the base, neutral, and acid compounds, and the spiking solution must contain the representatives of all three groups.

Example 4.17:   **Spiking solution composition**

VOC analysis by EPA Method 8260
(may be also used for EPA Method 8021)
- 1,1-Dichloroethene
- Trichloroethene
- Chlorobenzene
- Toluene
- Benzene

Organochlorine pesticides
by EPA Method 8081
- Lindane
- Heptachlor
- Aldrin
- Dieldrin
- Endrin
- DDT

SVOC analysis by EPA Method 8270

Base/Neutral Fractions
- 1,2,4-Trichlorobenzene
- Acenaphthene
- 2,4-Dinitrotoluene
- Pyrene
- *N*-Nitroso-di-*n*-propylamine
- 1,4-Dichlorobenzene

Acid Fraction
- Pentachlorophenol
- Phenol
- 2-Chlorophenol
- 4-Chloro-3-methylphenol
- 4-Nitrophenol

Laboratories select the LCS/LCSD concentration either at a level recommended by the EPA (EPA, 1996a) or at their own discretion. The EPA-recommended spiking levels include the following choices:

- Regulatory action level
- One to five times above the background concentration
- Twenty times the PQL
- Near the midrange calibration point

For a production laboratory, the most practical choice is a concentration near the midrange calibration point, as it does not require dilutions, usually has a good recovery, and brings consistency into daily batch preparation.

After the recovery of the spiked compounds and the RPD between the LCS and LCSD recoveries have been calculated (Equations 1 and 2, Table 2.2), the laboratories evaluate them statistically using a QC tool known as ***control charts***. These are the graphical representations of statistically evaluated data that enable laboratories to maintain the analytical process in the state of statistical control and to detect unwanted trends or deviations. Control charts may be substituted with databases that contain the information on values, control limits, and trends in a numerical, instead of a graphical, form.

An example of a recovery control chart is shown in Figure 4.7. The mean recovery of individual measurements is represented by the centreline. The *upper warning limit* (UWL) and the *lower warning limit* (LWL) are calculated as plus/minus two standard deviations (mean recovery $\pm 2s$) and correspond to a statistical confidence interval of 95 percent. The *upper control limit* (UCL) and the *lower control limit* (LCL) are calculated as plus/minus three standard deviations (mean recovery $\pm 3s$), and represent a statistical confidence interval of 99 percent. Control limits vary from laboratory to laboratory as they depend on the analytical procedure and the skill of the analysts.

For the analytical process to be in control, the LCS and LCSD recoveries must fall within the UCL and the LCL. If a recovery falls outside these limits, the laboratory must take corrective action, such as instrument recalibration or sample re-extraction and reanalysis.

The EPA recommends that the control limits of 70–130 percent be used as interim acceptance criteria until the laboratory develops its own limits (EPA, 1996a). In reality, not every laboratory evaluates their analytical precision and accuracy statistically; many rely on the EPA guidance or choose control limits arbitrarily. The typical arbitrary control limits are 50–150 percent; these limits are sufficiently wide to encompass the recoveries of most organic analytes. As a rule, the control limits of 65–135 percent reflect the typical laboratory accuracy for most organic analytes; for metals these limits are 75–125 percent. The arbitrary control limits do not reflect the actual laboratory performance and their routine use is an unacceptable laboratory

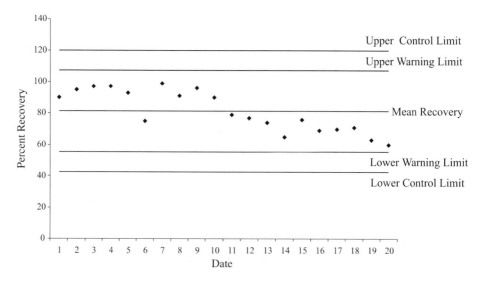

Figure 4.7   Example of a recovery control chart.

practice. The absence of statistical control limits is a serious deficiency of the laboratory quality system. *A laboratory that does not maintain statistical control charts does not know the accuracy and precision of its own analysis.*

Due to poor laboratory technique, statistically determined limits may be inadequately wide. For example, recovery limits of zero to 200 percent are not acceptable for any analyte, even if determined statistically; they indicate only a poor quality of laboratory work. Such control limits should not satisfy a data user who is concerned with data quality.

4.6.1.4 Matrix spikes

Matrix spikes are field samples spiked with known concentrations of target analytes and carried throughout the entire preparation and analysis. The MS/MSD spiking solutions are usually the same as used for the preparation of LCS/LCSD.

Matrix spikes enable the data user to evaluate the extent of matrix interference effects on the recovery of target analytes and to draw conclusions on the accuracy of environmental sample data. Soils of certain lithological composition have a tendency to retain chemicals, a property that is often demonstrated by MS/MSD recovery data. Such information may be particularly important for the projects where sample concentrations are compared to action levels. Low recoveries from certain soil types may render results inconclusive and warrant a different sampling or analysis approach. That is why *matrix spikes provide meaningful data only if performed on the project samples.*

An MS/MSD pair also provides the precision data that may be used in data quality evaluation. These data, however, may be affected by the variability of contaminant distribution in the matrix and for this reason may not be as useful as the LCS/LCSD data.

We should collect additional sample volumes and instruct the laboratory to use them for *at least one* MS analysis for every 20 field samples, especially for soil from uncharacterized sites. Otherwise, the laboratory may furnish the data on a randomly selected batch matrix spike prepared on a sample collected for a different project. Obviously, this batch matrix spike will not provide any useful information on matrix effects at another site.

4.6.1.5 Laboratory duplicates

Laboratory duplicates are field samples that are prepared and analyzed twice. Laboratories subsample two aliquots of a field sample from the *same container* and treat them as two separate samples. Typically, laboratory duplicates are prepared and analyzed with every 10 samples in trace element and inorganic analyses. This allows laboratories to calculate the RPD between the two results as a measure of analytical precision. Laboratory duplicates are rarely used in organic compound analysis.

4.6.1.6 Surrogate standards

Surrogate standards are organic compounds added to field samples and laboratory QC samples prior to preparation; they are similar in chemical behavior to the target analytes and are not expected to be present in samples. The laboratory must demonstrate that the surrogate standards do not interfere with target analytes.

Surrogate standard recovery measures analytical accuracy for *each individual sample*. Approved methods for organic compound analysis usually recommend the surrogate standard selection. Similar to target analytes, multipoint calibration curves are prepared for surrogate standards.

The importance of surrogate standard recovery in evaluating data quality cannot be overemphasized. Laboratories evaluate the efficacy of extraction of each individual sample based on the surrogate standard recovery. If batch QC checks are acceptable, but the individual sample surrogate standard recovery is not, the validity of sample results is questionable. *Results of organic compound analysis performed without surrogate standards cannot be considered definitive.*

Similar to LCS recoveries, surrogate standard recoveries should be monitored by the laboratory and plotted as control charts. The EPA recommends the use of in-house laboratory control limits for surrogate standards recoveries for all organic compound analyses (EPA, 1996a). The exception is the CLP SOW, which specifies these limits for soil and water analysis. Unless affected by matrix interferences, surrogate standard recoveries normally have relatively narrow control limits, 65–135 percent for most organic compound analysis. (Many laboratories, however, default to arbitrary limits of 50–150 percent for GC analyses, instead of using statistical control limits.)

For GC/MS analyses, some laboratories use surrogate standard recovery limits from outdated versions of EPA Methods 8260 and 8270. These recovery limits, shown in Example 4.18, are fairly close to the statistically derived control limits at most laboratories and can be safely used in the evaluation of data quality. The surrogate

---

Example 4.18:   Surrogate standard recovery limits for GC/MS methods

These surrogate standard recoveries are from the CLP SOW (EPA, 1999d). They are also acceptable for EPA Method 8260 and 8270.

Surrogate Standard	Percent recovery, water	Percent recovery, soil
VOCs Analysis		
4-Bromofluorobenzene	86–115	74–121
Dibromofluoromethane	86–118	80–120
Toluene-d_8	88–110	81–117
1, 2-Dichloroethane-d_4	80–120	80–120
SVOC Analysis		
Base/Neutral Fractions		
Nitrobenzene-d_5	35–114	23–120
2-Fluorobiphenyl	43–116	30–115
Terphenyl-d_{14}	33–141	18–137
1, 2-Dichlorobenzene-d_4	16–110	20–130
Acid Fraction		
Phenol-d_5	10–94	24–113
2-Fluorophenol	21–100	25–121
2, 4, 6-Tribromophenol	10–123	19–122
2-Chlorophenol-d_4	33–110	20–130

standards used in EPA Method 8270 include base, neutral, and acid compounds to better estimate the extraction efficacy of the three target analyte classes.

4.6.1.7 Internal standards

Internal standards are brominated, fluorinated or stable isotopically labelled analogs of specific target compounds or other closely related compounds not found in environmental samples that are added to all standards, field, and laboratory QC samples as part of internal standard calibration procedure. The addition of internal standards takes place *after* the samples have been prepared and *before* they are analyzed.

As individual sample QC checks, internal standards are important in compound quantitation. They should be monitored with the same care as other QC checks. The deterioration of internal standard area counts (the area under a chromatographic peak) reflects the changes in the analytical system that may eventually degrade the quality of analysis to an unacceptable level. EPA Methods 8260 and 8270 require that the area for each internal standard be within the range of -50 to $+100$ percent of the area of the internal standards in the most recent CCV (EPA, 1996a). This requirement may be used as a general rule for all other methods that use internal standard calibration.

4.6.2 Initial demonstration of capability

As part of the QC program, laboratories must conduct initial demonstration of capability that includes method validation studies and operator competence certification.

The interpretation and implementation of published methods invariably differ at different laboratories due to diversity of utilized instruments, their incidental elements and supplies, and the differences in method interpretation. Each analytical method must be validated at the laboratory before it is used for sample analysis in order to demonstrate the laboratory's ability to consistently produce data of known accuracy and precision. Method validation includes the construction of a calibration curve that meets the acceptance criteria; the determination of the method's accuracy and precision; and the MDL study. A method SOPs must be prepared and approved for use. Method validation documentation is kept on file and should be always available to the client upon request.

Each analyst performing analysis must be trained in the procedure, and the training must be documented in the laboratory records. The training establishes single analyst precision through the preparation and analysis of a blank and four LCSs. (Single analyst precision is the *repeatability* or *intralaboratory precision*, a secondary DQI.)

4.6.3 System and performance audits

The performance of the laboratory QC system is assessed through internal and external systems audits. Laboratory QA section performs *internal audits* and identifies the weaknesses of the quality system or the deviations from approved internal procedures. The state, the EPA, a client or any other body that oversees the quality of laboratory work conducts *external audits* as part of initial and ongoing laboratory accreditation process. To verify laboratory compliance with the requirements of a particular SOP, a method, or a SAP, an internal or external *compliance audit* may be conducted.

Laboratory clients use audits for prequalification of analytical laboratories and as a quality assessment tool during project implementation. Recurring laboratory problems related to data quality often warrant an audit from the client to help the laboratory identify and correct the causes.

Only experienced quality auditors, such as the chemists who are knowledgeable in the analysis being audited, should conduct external audits. Audit findings give rise to corrective action that is implemented and documented by laboratory operations personnel. However, even the most detailed and frequent audits may be ineffective in disclosing data fraud (Popek, 1998b). External audits are expensive and are usually conducted for government project or major industry client work.

The data user may assess laboratory performance through analysis of *single-blind* and *double-blind PE samples* for project-specific analytes. Single-blind samples are clearly identified as PE samples, whereas double-blind samples are disguised as field samples. Certified blind PE samples may be obtained from several manufacturers in the USA. They are prepared as water and soil samples spiked with a variety of common contaminants. The manufacturer collects the interlaboratory recovery data for these samples and calculates acceptance criteria that are available to the PE sample purchaser.

Recovery data are usually not available for custom-made PE samples with contaminant concentrations prepared per client request. PE samples without acceptance criteria are not as useful as the ones with statistically derived acceptance criteria. The data obtained from such samples reflects the laboratory's ability to correctly identify contaminants, rather than to accurately quantify them.

From a perspective of the data user, a combination of audits and PE samples may be a cost-effective approach to laboratory quality assessment. Single-blind and double-blind PE samples are typically used on long-term, large sample volume projects. Because of the additional costs, they are not sent to the laboratory with every shipment of field samples. PE samples for the project contaminants of concern analyzed at the start of project activities will enable the data user to establish the laboratory's performance. Depending on the project duration and on the volume and importance of the collected samples, additional PE samples may be analyzed on a quarterly or semiannual basis.

EPA 'round robin' studies is another quality assessment tool that is effective in revealing deficiencies in laboratory's ability to correctly identify and quantify environmental pollutants. The EPA performs these studies on a regular schedule and evaluates the results for the purpose of establishing interlaboratory precision, which is a measure of variability among the results obtained for the same sample at different laboratories. (*Interlaboratory precision* or *reproducibility* is another secondary DQI.)

Practical Tips: Laboratory quality assurance

1. Obtain control limits from the laboratory that will be providing analytical services to the project and include them in the SAP.

2. Examine results of the recent EPA 'round robin' study as they indicate the laboratory's ability to perform accurate analysis. A laboratory that does not participate in the EPA studies cannot be compared to other laboratories in terms of analytical accuracy.

3. Perform a prequalification audit of the laboratories that will provide analytical services for your projects. If an audit is not feasible due to budget or time constraints, request a copy of the latest audit report that was conducted at the laboratory by an accrediting agency. As a rule, a frequently audited laboratory is a better laboratory.

4. In case there are doubts in data quality, request from the laboratory the SOPs, the MDL studies, and method validation documentation. Have them examined by a qualified chemist to determine if they are acceptable.

To reduce the data validation expense, a combination of Level 4 and Level 3 validation may be used without compromising data quality. For example, 20 percent of the project data may undergo a Level 4 validation and 80 percent will then have a Level 3 validation. A 10 percent Level 4 and 90 percent Level 3 combination is often considered sufficient for many decisions. Example 5.1 illustrates several validation strategies that may be used for projects with different objectives.

Example 5.1: Data validation strategies

Confirmation sampling for site closure

- Soil contaminated with hazardous materials is being excavated from a site, and confirmation samples are collected to verify that the cleanup levels have been met. The SAP specifies a 20 percent Level 4 and an 80 percent Level 3 data validation of the planned 120 samples. The chemist coordinates the sampling efforts and as the samples are being collected identifies a COC Form with 15 samples and a COC form with 10 samples. These two COC Forms represent approximately 20 percent of the total number of samples. The chemist notifies the laboratory that these two sample groups will require Level 4 data packages. The rest of the samples will require Level 3 data packages. The chemist states the requirements for the data package content on every COC Form.
- On the same project, samples of excavated contaminated soil will be collected for disposal profiling. The analytical data will undergo only a cursory review. The SAP specifies standard laboratory packages for these samples. The chemist verifies that this requirement is stated on the COC Forms before the samples are shipped to the laboratory.

Site investigation

- Three hundred and sixty samples are being collected during a site investigation project. The collected data will be validated at 10 percent Level 4 and 90 percent Level 3. The chemist reviews the COC Forms for the samples on the days when samples are being collected and identifies three COC Forms, containing 14 samples each. The chemist then requests Level 4 data packages for these 3 groups of samples (representing approximately 12 percent of all samples) and Level 3 data packages for the rest of the samples.

Data qualifiers

In the course of data validation, data qualifiers are attached to the data. ***Data qualifiers are the alphabetic symbols that indentify an undetected compound or a deviation from acceptance criteria***. Data qualifiers are also called ***data flags***. The findings of data validation are detailed in a data validation report, which documents the validation process and explains the reasons for attaching the qualifiers to the data. Laboratories also use data qualifiers for indicating deviations from laboratory acceptance criteria. These qualifiers are replaced with the validation qualifiers in the course of data validation. Qualifiers are rarely used in data review.

A great variety of data qualifiers, each signifying a specific deviation from acceptance criteria, may be used in data validation. The trends in deviation type and incidence may be important for identifying systematic field and laboratory errors. However, they are not as important to the data user who only needs to know whether

Table 5.2   Data validation qualifiers and their use

Qualifier	Use
J	1. The compound was positively identified but the concentration is an estimated quantity, which is greater than the method detection limit and lower than the reporting limit. 2. The compound was positively identified at a concentration above the reporting limit, however, the quantitation is not definitive. The concentration is an estimated quantity.
U	The compound was analyzed for, but not detected at a concentration above the reporting limit. 'U' is attached to the reporting limit numerical value to indicate that the compound was not detected.
R	The data are rejected because the associated laboratory QC checks failed to meet acceptance criteria. 'R' is attached to the data that are not valid within the framework of the applied standard.
UJ	The compound was analyzed for, but was not detected at a concentration above the reporting limit. The associated reporting limit value is an estimate due to deficiencies in laboratory QC.

the data are valid or not. Deviations from various acceptance criteria ultimately place the data into three categories:

1. Definitive—valid data that can be used for project decisions
2. Estimated—valid data that can be used for project decisions
3. Rejected—invalid data that cannot be used for project decisions

The four common qualifiers shown in Table 5.2 are sufficient for identifying these three categories.

Data qualified as estimated values are used in the same manner as valid, unqualified values for making project decisions, in completeness calculations, and in statistical calculations. After all, all environmental data are only the approximations of true contaminant concentrations.

5.2   THE SEVEN STEPS OF DATA EVALUATION

Validation and review processes, different as they may be in the level of diligence, follow the same basic steps of consecutive examination of a data package to establish the following facts:

- Acceptance criteria for data quality indicators have been met
- The data are technically and legally defensible
- The data are representative of the conditions at the sampling point at the time of sampling

This examination may be broken down into seven steps shown in Figure 5.2, each addressing a key aspect of data quality.

Each data package representing a group of samples analyzed with one method is evaluated separately. The end product of data evaluation is an assessment of several

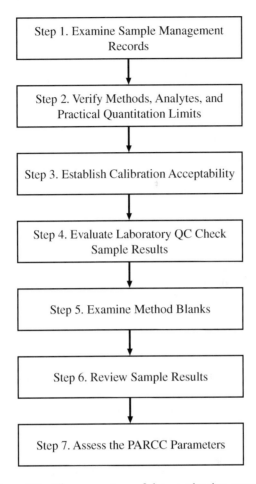

Figure 5.2 The seven steps of data evaluation process.

PARCC parameters (typically, accuracy, precision, and representativeness) for the samples in the reviewed data package. A checklist prompts a structured approach to data evaluation and allows documenting deficiencies. Appendices 23, 24, and 25 contain examples of such checklists for organic, trace element, and inorganic analyses.

In the course of data evaluation, the chemist will make decisions on data quality by relying not only on acceptance criteria but also using the knowledge of physical and chemical properties of the target analytes and their fate in the sampled matrix. The chemist's professional expertise and judgment play a major role in data evaluation.

This chapter describes the process of data evaluation that is equivalent to data validation Level 3. It may be scaled up to validation Level 4 by including a thorough examination of raw data and the recalculation of results or scaled down to cursory data review.

5.2.1 Step 1—Examine sample management records

Data evaluation process starts with a review of the Case Narrative prepared by the laboratory as part of Level 3 and 4 data packages. The Case Narrative summarizes errors and omissions in sample handling and laboratory QC deficiencies; describes matrix abnormalities that were identified in the course of analysis and internal laboratory review; and explains the use of laboratory data qualifiers associated with these events. The Case Narrative does not address the issues of data validity or usability and is merely an enumeration of laboratory events that were out of compliance with laboratory acceptance criteria. The decisions on data validity and usability will be made in the course of data quality assessment based on the intended use of the data.

The chemist also examines sample management records, such as the COC Forms, Laboratory Cooler Receipt Forms, and project records of communications with the laboratory to establish whether errors may have compromised the representativeness or legal defensibility of the samples. Review of sample management records enables the chemist to answer the following questions:

- Is the COC Form signed by all relinquishing and receiving parties?
- Were custody seals, if used, intact when samples were received at the laboratory?
- Were the samples collected into appropriate containers?
- Were the samples properly preserved?
- Was a temperature blank enclosed in the cooler?
- Was the temperature of the cooler within the range of 2–6°C?
- Did all of the sample containers reach the laboratory intact (unbroken)?
- Was a trip blank included with the samples for VOC analysis?

Positive answers to all of these questions establish the 'collected sample' *representativeness* (based on proper preservation, containers, and shipping), and the legal defensibility of the sample shipment (the samples were not tampered with and have a documented custody chain).

Samples with compromised integrity may lose the property of representativeness or become legally vulnerable. Common causes of compromised sample integrity are the improper preservation, incorrect or broken sample containers, and elevated cooler temperature. Example 5.2 shows how professional judgment helps in making decisions on data quality for samples with questionable integrity.

5.2.2 Step 2—Verify methods, analytes, and practical quantitation limits

Verifying that appropriate methods have been performed by the laboratory is an important part of data evaluation. A common laboratory error is the substitution of requested analytical methods with parallel methods from a different environmental regulation series. Comparable regulatory methods may have different target analyte lists, PQLs, calibration requirements, as shown in Tables 4.5, 4.6, and 4.7, and QC checks. In addition, laboratories perform the published methods according to their own procedures, and these procedures, may or may not meet the QC requirements of a specific regulatory method. Data comparability evaluation, if planned for the project, may also be affected by the differences in the preparation and analysis methods. In Step 2, the chemist identifies the procedure and methodology errors by focusing on the following issues:

Example 5.4:   Use acceptance criteria based on laboratory control limits

The SAP establishes the advisory accuracy acceptance criteria for EPA Method 8081 of 70–130 percent. The laboratory control limits for accuracy exceed the SAP specification as follows:

Contaminant of concern	*Laboratory control limits or accuracy as percent recovery*
Alpha-BHC	26–120
Delta-BHC	47–123
Gamma-BHC	23–150
Endosulfan II	49–169
Methoxychlor	48–137
Endrin Aldehyde	20–165

The chemist understands that advisory acceptance limits stated in the SAP do not reflect the actual accuracy of analysis. The laboratory has determined the accuracy by using control charts for LCS data. The QC check sample data that are within these statistical limits indicate that the analytical process is in control. The chemist accepts the laboratory accuracy criteria and does not qualify the sample results, although the SAP acceptance criteria for LCS/LCSD have been exceeded.

---

quality. If the actual laboratory control limits are not available in the planning phase, the SAP should have a provision for replacing arbitrarily selected or provisional advisory limits with actual laboratory limits in the data evaluation process. Example 5.4 underscores this point.

If project-specific matrix spike data are available, the chemist evaluates them to establish the effects of matrix interferences on the accuracy and precision of project sample analysis. Similar to LCS/LCSD recoveries, MS/MSD recoveries are monitored at the laboratory as control charts; these recoveries, however, are not used as acceptance criteria for qualifying the data for the whole batch. The RPD between the MS and MSD results is an additional measure of analytical precision that may be used when the LCS/LCSD precision has failed or is not available.

5.2.5   Step 5—Examine method blanks

Contaminated samples and method blanks and instrument memory effects (carry-over) are a major source of false positive results. To determine whether data interpretation errors may have produced false positive results, the chemist examines the instrument and method blank data and answers the following questions:

1. Was a method blank prepared and analyzed?
2. Do instrument blanks indicate the presence of memory effects?
3. Is the method blank contaminated?
4. If contaminants are present in the instrument and method blanks, are they among the project's contaminants of concern?
5. Are the concentrations of contaminants of concern in the method blank close to the concentrations in the samples?
6. Are results of interference check samples in ICP–AES analysis acceptable?
7. Is the frequency of instrument and method blank analysis acceptable?

Table 5.4 summarizes the acceptance criteria for instrument, calibration, and method blanks. No contaminants of concern should be present in method blanks above the laboratory PQL. Equally important is that instrument blanks show no memory effects. If these conditions are not met, a possibility for false positive sample results becomes real. For decision on sample data with contaminated method blanks the chemist may rely on the following rules of the Functional Guidelines:

Rule 1
Sample results for *common laboratory contaminants* are qualified by elevating the PQL to the concentration found in the sample if sample concentration is less than 10 times the blank concentration. These contaminants include methylene chloride, acetone, 2-butanone, cyclohexane in VOC analysis and phthalates in SVOC analysis.

Rule 2
In organic analysis, results for any compound detected in the sample (rather than the common laboratory contaminants) are qualified by elevating the PQL to the concentration found in the sample if the sample concentration is less than five times the blank concentration.

Rule 3
For trace element analysis, if the preparation blank contains target analytes above the PQL, the lowest concentration of that analyte in a sample must be 10 times the blank concentration for a result to be acceptable. Samples with concentrations below 10 times the blank concentration must be redigested and reanalyzed with a new preparation blank.

Rule 4
If the blank concentration is higher than the PQL and the sample concentration is lower than the PQL, the sample result is reported as undetected at the PQL.

Rule 5
Results of organic and inorganic analyses must not be corrected by blank subtraction.

The application of these rules is illustrated in Example 5.5.

The chemist will pay particular attention to samples, which were diluted in the course of analysis. The concentrations of laboratory contaminants that may be present in the samples, as indicated by contaminated instrument or method blanks, will be magnified by the multiplication by the dilution factor and create false positive results of high concentration values.

5.2.6 Step 6—Review sample results
In Steps 1 through 6 the chemist has evaluated batch QC check results and came to a conclusion on the acceptability of the whole batch. In Step 6 the chemist focuses on individual sample QC checks and on remaining qualitative and quantitative aspects of analysis. The chemist reviews the data for each individual sample in order to evaluate the following issues:

Example 5.5: Recognize and eliminate false positive results

Analyte	PQL	Method blank result	Sample result	Rule applied	Qualified result
Methylene chloride	10	25	220	1	220U (Undetected)
Acetone	5	15	160	1	160 (No qualification)
Benzene	1	3	10	2	10U (Undetected)
Toluene	1	5	27	2	27 (No qualification)
Zinc	10	30	50	3	50R (Rejected)
Lead	5	30	350	3	350 (No qualification)
Di-*n*-octylphthalate	5	10	4	4	5U (Undetected)
Zinc	10	30	8	4	10U (Undetected)

- Was each sample prepared and analyzed within the method-required holding time?
- Are surrogate standard recoveries in each sample acceptable?
- Did matrix interferences or sample dilutions affect the surrogate standard recoveries and reporting limits?
- Are the peak areas of internal standards in GC/MS analysis acceptable?
- Was second column confirmation, if required by the method, performed for individual compound analysis with GC methods?
- Do results from both columns agree?
- Was the post-digestion spike addition performed?
- Are results of post-digestion spike recoveries acceptable?
- Were serial dilutions performed?
- Are results of serial dilutions acceptable?
- Was the MSA performed to overcome the effects of matrix interferences in elemental analysis?
- Was moisture correction, if required, performed for the final results?
- Do final results for contaminants of concern fall within the calibration range?

Deviations from method-required holding time may affect the collected sample representativeness. Depending on the severity of a deviation, the chemist may qualify the data as estimated values or reject them as invalid. Holding time violations have a different effect on volatile and semivolatile compounds and that is why the chemist's professional judgment is particularly important in such situations.

GC/MS methods are the only published methods that include the surrogate standard recovery limit guidance. Similar to LCS, acceptance criteria for surrogate standard recoveries of all other organic analysis methods are the laboratory control limits. The limits for internal standard recovery in GC/MS analysis are specified by the method and cannot be changed by the laboratory. Acceptance criteria for matrix interference detection techniques in trace element analyses, discussed in Chapter 4.4.4.5, are also specified in the analytical methods.

The chemist reviews results of each analysis and determines whether data qualification is needed. Typical deficiencies that turn definitive data points into estimated ones include insufficient surrogate standard recoveries; the absence of second column confirmation; and the quantitation performed outside the calibration curve. The chemist may even reject the data based on low surrogate standard recoveries. Example 5.6 shows how surrogate standard recoveries may affect the validity of analytical results.

Example 5.6:   Examine surrogate standard recoveries

The laboratory control limits of the BFB surrogate standard in EPA Method 8021 are 56 to 143 percent. The chemist may use the following rationale for qualifying individual sample data with the surrogate standard recoveries outside these limits:

	Percent recovery			
	56 to 143	> 143	10 to 56	<10
Detected analytes	Not qualified	Estimated (J)	Estimated (J)	Estimated (J)
Undetected analytes	Not qualified	Not qualified	Estimated (UJ)	Rejected (R)

The surrogate standard recoveries for some samples may be outside the control limits due to sample dilutions or matrix interferences. Such samples are usually clearly identified in the Case Narrative or in the laboratory reports, and their data are not qualified. For example, if a surrogate standard recovery is below 10 percent due to dilution, the result will not be rejected and will not be qualified as an estimated value. The chemist, however, may request from the laboratory the raw data for this sample in order to verify whether the dilution was justified and the interferences were truly present.

5.2.7   Step 7—Assess the PARCC parameters

After having completed the first six steps of the evaluation process, the chemist has sufficient information to make a statement on *precision, accuracy,* and *certain aspects of representativeness* of the evaluated samples. The **completeness** of the data set may be also assessed at this step, if the data for all of the samples collected for a specific intended use are in the evaluated package. If there is more than one data package, then completeness will be assessed later in the DQA process, after the packages for all samples have been evaluated.

By summarizing the evaluation findings in a written statement, the chemist will answer the following questions:

- Is the collected sample representativeness uncompromised?
- Are analytical accuracy and precision within the acceptance criteria specified in the SAP?
- Is the completeness goal specified in the SAP met? (optional at this step)

For sample results with complete and thoughtfully compiled data packages, these questions may be answered immediately upon evaluation. However, if QC check data and support documentation are missing or are inaccurate due to data management errors at the laboratory, the chemist's decision on data quality may be delayed. The chemist will request that the laboratory provide additional data in order to evaluate them at a later date. The loss of continuity in the data evaluation process due to poor quality of data packages is counterproductive and may cause delays in the scheduled project report delivery to the client.

Unqualified data and estimated values are considered valid and can be used for project decisions, whereas the data points, which were rejected due to serious deficiencies in representativeness, accuracy, or precision, cannot. In an attempt to obtain data of better quality, the chemist may ask the laboratory to reanalyze some of the samples or extracts, if they are still available and have not exceeded the holding time. Depending on the number of the rejected data points and their importance for project decision, the chemist may recommend resampling and reanalysis.

5.2.8   Validation *versus* review
The above-described data evaluation is a relatively detailed process, which is equivalent to Level 3 validation. Despite its thoroughness, it provides the data user with only *partial knowledge* of data quality. This partial knowledge, however, is considered to be sufficient for making decisions for various types of environmental projects listed in Table 5.1.

Level 4 validation
A *complete knowledge* of the data quality that arises only from Level 4 validation enables the data user to make project decisions with the highest level of confidence in the data quality. That is why Level 4 validation is usually conducted for the data collected to support decisions related to human health. Level 4 validation allows the reconstruction of the entire laboratory data acquisition process. It exposes errors that cannot be detected during Level 3 validation, the most critical of which are data interpretation errors and data management errors, such as incorrect computer algorithms.

The process of Level 4 data validation follows the same basic steps as Level 3 validation, although the evaluation is much more detailed and, consequently, more time-consuming. The following additional review items are included into Level 4 validation:

- Target compound identification (from spectra, chromatograms)
- Recalculation of analytical results
- Review of tentatively identified compounds
- Review of preparation bench sheets
- Review of raw data for false negative results (unreported compounds that may be present in the samples)

Cursory review
On the opposite end of the data evaluation range is cursory data review. It is an abbreviated evaluation process in which Step 3, 'Establish Calibration Acceptability,'

is omitted. Calibration data are not evaluated during cursory review and they are not included in the data package. Data are not qualified based on cursory review, and because the amount of valid data is not established, completeness is not calculated. Cursory review makes the quality of data known only within the limitations of the provided information, and the critical issues of calibration acceptability, compound identification, and quantitation remain unknown.

Cursory review provides the data user with a *limited knowledge* of data quality. It is appropriate for the determinations of gross contaminant concentrations for certain project tasks as shown in Table 1. The use of the data that have undergone only a cursory review for making decisions with the consequences for human or ecological receptors would be a serious mistake.

Practical Tips: Data evaluation

1. Data validation is a specialized process, best conducted by data validation professionals. Hiring a validation company for data validation Level 4 and Level 3 is a cost-effective option.

2. A well-prepared Case Narrative will instantly alert the chemist to deficiencies in sample handling and analysis.

3. As a rule, if the cooler temperature is between 6 and 10°C, the data do not have to be qualified.

4. Discharge permit requirements specify the analytical methods to be used. The oversight agencies typically will not accept results of comparable analytical methods from a different environmental regulation series.

5. The most significant consequences of method substitution at the laboratory are incomplete target analyte lists and the PQLs that may be different from the SAP specifications.

6. TPH analyses are particularly prone to false positive results. False positive GRO results are often reported for samples with a single peak, typically a solvent, present in the chromatogram in the absence of a gasoline pattern. Memory effects (carryover in the chromatographic column) often produce false positive results in TPH analysis.

7. A surrogate standard recovery that is above the upper control limit is not necessarily an indication of a high bias in the sample result. It may indicate matrix interference, a preparation error, or a measurement error.

5.3 THE SEVEN STEPS OF THE DATA QUALITY ASSESSMENT

The EPA developed a document titled *Guidance for Data Quality Assessment, Practical Methods for Data Analysis, EPA QA/G-9* (EPA, 1997a) as a tool for project teams for assessing the type, quality, and quantity of data collected for projects under the EPA oversight. This document summarizes a variety of statistical analysis techniques and is used primarily by statisticians. DQA, however, is not just a statistical evaluation of the collected data. It is a broad assessment of the data in the context of the project DQOs and the intended use of the data, which requires a substantial insight into laboratory QA/QC procedures and the knowledge of organic, inorganic, and analytical chemistry that statisticians may not have. A chemist, experienced in project planning, DQO development, field sampling, and laboratory analysis is the best candidate for conducting DQA and engaging a statistician, if necessary.

From the perspective of the chemist, DQA can be broken down into seven steps shown in Figure 5.3. Similar to the DQO process, the steps may undergo several iterations before the final statement is formulated; depending the purpose of data collection, some of these steps may be modified or even omitted. For example, a DQA conducted for hazardous waste disposal data will not be nearly as extensive as the one for a risk assessment project.

The DQA Steps 1 and 6 address the issue of data *relevancy*, whereas Steps 2, 3, 4, and 5 establish data *validity*. A summary statement on the relevancy and validity of the data is prepared in Step 7.

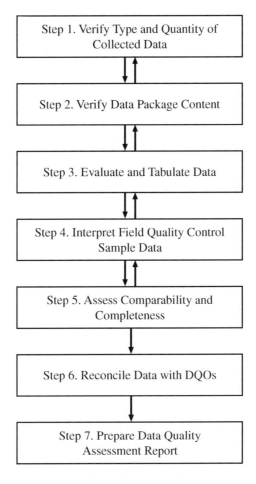

Figure 5.3   The seven steps of data quality assessment.

A stepwise approach to DQA identifies different tasks that may be performed by individuals with different expertise. For example, a less experienced chemist may verify the data package content (Step 2), whereas a more experienced chemist may perform data evaluation (Step 3). For a statistical data collection design, a statistician may be involved in the assessment of data relevancy (Step 6). A database manager may be involved at several steps if the EDDs are part of laboratory deliverables and if completeness is calculated.

5.3.1   Step 1—Verify type and quantity of collected data

Step 1 of DQA starts when the project is still in the field and while samples are being collected and submitted to the laboratory for analysis. Ongoing verification of the *types* and *quantities* of the data being collected is one of the key elements of DQA. It helps us to assure that the answer to the question '*Do the data have a well-defined need, use, and purpose?*' is affirmative.

Although the procedures for the collection of relevant data have been specified in the SAP, field implementation is not an error-free process; in addition, the changes in field conditions often cause deviations from the planned action. A real-time oversight is necessary for assuring that field crews are implementing the provisions of the SAP properly; errors, if made, are immediately corrected; and that field changes are justified, documented, and not contrary to the project DQOs.

While field crews collect the samples, the chemist verifies their types and quantities by comparing the COC Forms, Sample Tracking Logs, and other field records to the SAP specifications. The chemist interacts with the field crews for the resolution of errors in the sampling point selection, sample containers and preservation or COC Form errors. The chemist also serves as a point of contact with the laboratory, if any technical issues arise during sample analysis.

The chemist reviews analytical results as they are received from the laboratory. It is particularly important for remediation projects with field decisions pending results of analysis. Laboratory data often guide remedial activities, such as soil removal, treatment plant operations, and compliance monitoring. For these projects, reliable data with expedited turnaround time of analysis may be necessary. To avoid making decisions on data of completely unknown quality, the chemist reviews such data in a cursory manner to verify that the basic laboratory QC requirements have been met, whereas a complete evaluation of data packages will be conducted later.

For projects with permitted discharges, the chemist may review analytical methods against permit requirements. Projects that operate under POTW or NPDES permits may have strict analytical requirements, and permit-required analytical methods cannot be substituted. If the laboratory has erroneously substituted a method, the obtained data valid as they are, may not be relevant and usable.

5.3.2   Step 2—Verify data package content

Data packages and EDDs of varying complexity that accompany the final analytical results are necessary for data evaluation. Although the requirements for data package content and the EDD format have been specified in the SAP and in the Laboratory SOW, data packages and the EDDs may be incomplete or incorrect. The chemist must verify the correctness of the supplied information shortly after the packages have been received from the laboratory, as only complete and correct data packages can be evaluated.

The verification does not take long, if the data package has been compiled according to the SAP specifications; however, it may become a lengthy matter, if the laboratory has made errors in the compilation and if new information is needed. The verification is most effective, if conducted immediately after the data packages have been received by the data user, because the laboratory has not yet had the time to archive the raw data and support documentation, such as laboratory records, magnetic tape, and electronic data.

EDD contents must be compared to the hardcopy results as errors may also take place in the EDD production. This task is usually performed with the assistance of the database manager. If discrepancies have been discovered, the laboratory will make the necessary corrections to the hardcopy and/or the EDD.

After the data packages have been verified to contain all of the required information, the data may be evaluated.

5.3.3   Step 3—Evaluate and tabulate data

Data evaluation that may be conducted at three levels of diligence, Level 4 validation, Level 3 validation, and cursory review, is described in Chapter 5.2. The end product of data validation is the qualified data and a statement related to *accuracy*, *precision*, and *representativeness* of a group of samples.

After data evaluation for all project samples has been completed, the chemist sorts the qualified results and the data evaluation statements according to the intended use of the data as illustrated in Example 5.7. *From this moment on, the data will be assessed based on their intended use.*

Example 5.7:   Tabulate data according to their intended use

Over the course of the project, the following samples were collected:

- Eighty-five soil samples from the bottom of the excavation
- Sixteen soil boring samples
- Four soil samples from a contaminated soil stockpile
- Eight groundwater samples from monitoring wells
- One wastewater sample from combined well purging and equipment decontamination wastewater streams

The data packages comprise results for soil and water samples according to the COC Form, regardless of the intended data use. The intended data use includes site confirmation, site characterization, groundwater monitoring, and waste profiling. The chemist prepares tabulated summaries of qualified data according to the intended data use as follows:

- Summary of excavation confirmation sample results compared to action levels
- Summary of site characterization soil boring sample results
- Summary of groundwater monitoring sample results

These summaries will become part of the DQA statement. The chemist reviews waste disposal data, but does not tabulate it. Based on results of analysis, waste streams are profiled and shipped to the disposal facility under appropriate manifests.

5.3.4 Step 4—Interpret field quality control data

The interpretation of field QC data and their effects on the quality of sample data is conducted in DQA Step 4. This task may be also conducted in Step 4 of the data evaluation process, particularly if one and the same chemist performs data evaluation and DQA. Regardless of when it is performed, it is a distinctive and important task.

Results of the field QC samples (trip blanks, field duplicates, equipment blanks, etc.) are part of the data packages and, together with field samples, they are evaluated to establish whether they are valid. But what is the significance of these data? Like the field samples, these data were collected for a well-defined need, use, and purpose. They provide part of the answer to the question: '*Are the data representative of the sampled matrix?*'

5.3.4.1 Trip and equipment blanks

The chemist interprets the results of trip and equipment blank analyses to identify sample management errors during sampling, sample handling, and decontamination procedures and to determine whether these errors may have affected the collected sample representativeness. The chemist qualifies the data according to the severity of the identified variances from the SAP specifications and may even reject some data points as unusable. Example 5.8 shows a logical approach to the interpretation of the trip and equipment blank data.

5.3.4.2 Field duplicates

The chemist also evaluates field duplicate data to assess the total sampling and analysis precision of water and the inherent sample variability of soil as a measure of the sampling point representativeness. From the Sample Tracking Log or other field records the chemist identifies field duplicate samples and evaluates the correlation between the results of the primary sample and the duplicate.

Water duplicates

For water samples, the RPD of field duplicate results (Equation 1, Table 2.2) is a measure of total sampling and analysis precision. The RPD values are compared to the acceptance criterion specified in the SAP. A common field duplicate precision acceptance criterion for water samples is 30 percent. Field duplicate results that do not agree within 30 percent may indicate several problems as follows:

1. Poor sampling technique—as the result, the collected samples may not be representative of the sampled well water.
2. Free-floating product or sheen present in the samples—the very nature of these samples prohibits the collection of a representative sample.
3. Insufficient analytical accuracy may be a cause of poor field duplicate precision. For example, low analytical accuracy that is typical for samples with low contaminant concentrations reduces analytical precision.

The chemist identifies the probable cause of inadequate field duplicate precision and forms an opinion on a corrective action that will allow avoiding the same

---

Example 5.8:   Trip and equipment blank logic

1. Is the trip or equipment blank contaminated?

   - No. A clean trip or equipment blank signifies proper sample handling and shipping or adequate equipment decontamination. The sample data are representative of the sampled matrix.
   - Yes. Go to 2.

2. If any contaminants are present in the trip or equipment blank, are they among the project's contaminants of concern?

   - If compounds that are *not contaminants of concern* are present in the blanks, but not in the samples, then they most likely came with the water used in the blank preparation.
   - If compounds that are *not contaminants of concern* are present in the blanks and in the samples, then field or laboratory contamination may have taken place. The sample data are nevertheless, representative of the sampled matrix.
   - If contaminants of concern are present in the blanks and in the samples, go to 3.

3. Are the concentrations of contaminants of concern, if present in the trip or equipment blank and in samples, comparable?

   - If the concentrations of contaminants of concern in blanks exceed the sample concentrations, then the sample data are not representative of the sampled matrix and are invalid.
   - If the concentrations of contaminants of concern in the samples are greater than the concentrations in any of the blanks by a factor of 5 (or 10 for common laboratory contaminants), then the sample data are representative of the sampled matrix and may be used.

---

problem in the future. The data, however, are never qualified based on poor field duplicate precision.

Soil duplicates

The evaluation of soil field duplicates is somewhat complicated because numeric acceptance criteria are not appropriate for soil. Affected mainly by sample variability, the RPD values for soil sample duplicates may vary vastly. The sampling method also affects soil field duplicate precision, as shown in Chapter 2.6.2. Because soil duplicate RPD cannot be controlled, the application of a standard, such as an acceptance criterion, for their evaluation is not reasonable or practical.

While evaluating soil field duplicate data, the chemist may calculate the RPD as a relative measure of sample variability; the values, however, should be carefully interpreted in qualitative terms. If the same contaminants are identified and their concentrations are of the same order of magnitude, the variability is deemed negligible. Great disparities in the types and concentrations of the detected contaminants indicate significant variability. Example 5.9 illustrates this point.

The sample data are never qualified based on field duplicate precision, and the information gained from the evaluation of soil duplicates usually does not have any effect on project decision-making. The only situation, when these data may

Example 5.9:   Interpretation of field duplicate soil sample results

This example presents the typical results of field duplicate sample analysis that can be expected for VOCs and SVOCs in soil. The wide range of RPD values emphasizes the point that the use of numeric acceptance criteria for assessment of field duplicate precision is not practical.

Contaminant	PQL	Sample Result	Field Duplicate Result	RPD (percent)	Sample type
GRO (mg/kg)	5	140	62	77	Collocated grab samples
Toluene (µg/kg)	0.02	0.04	0.2	133	Unhomogenized splits
DRO (mg/kg)	5	19	28	38	Homogenized splits
RRO (mg/kg)	100	69J	120	54	Homogenized splits
Fluorene (µg/kg)	20	43	20U	Not calculable	Homogenized splits
Pyrene (µg/kg)	20	75	36	70	Homogenized splits
Lead (mg/kg)	0.2	20	160	156	Homogenized splits

Collocated grab sample and unhomogenized split data interpretation

- The RPDs are high, indicating high matrix variability. Because these samples cannot be homogenized due to contaminant volatility, high RPD values are expected.

Homogenized split sample data interpretation

- Homogenized duplicates collected for SVOC analysis show a wide range of RPD values. The detected concentrations are relatively low and are of the same order of magnitude. Analytical accuracy in the low concentration range is also low, affecting the precision of the two measurements. The data indicate an adequately performed homogenization procedure and have the precision that is normal for the detected range of concentrations.
- The lead concentrations are substantially higher than the PQL and vary significantly, indicating high matrix variability even in homogenized split samples.

occasionally affect project decisions, is site investigation projects, where high soil matrix variability may be considered an important factor in the development of the future sampling designs.

Poor correlation between field duplicate results is expected for certain types of samples and contaminants, as shown in Example 2.5, Chapter 2.6.2. For such samples, to reduce the cost of sampling and analysis, field duplicates should not be collected at all, as the data will not provide any meaningful information.

error (α), false acceptance decision error (β), and the gray range selected in Step 6 of the DQO process will be used in the statistical calculations. The outcome of the test will determine the choice between the null and the alternative hypotheses, which were formulated in the DQO process, with a stated level of confidence.

Practicing professionals, such as geologists, engineers, and chemists, typically do not have the expertise to conduct statistical data analysis. Only, experienced statisticians should perform complex statistical evaluations of environmental data. There is, however, one type of statistical calculations routinely performed by many, regardless of their professional background. It is the calculation of the confidence interval of the mean for the determination of the attainment of the action level using a *one-sample Student's t-test*. (*One-sample* in this context means one population mean.) This statistical test is detailed in several EPA guidance documents (EPA, 1996a; EPA 1997a).

The underlying assumptions of the Student's t-test include simple random and systematic sampling and a normal distribution of the sample mean. The upper limit of the confidence interval for the mean concentration is compared to the action level to determine whether solid waste contains a contaminant of concern at a hazardous level. (The calculation is conducted according to Equation 10, Appendix 1.) A contaminant of concern is not considered to be present at a hazardous level, if the upper limit of the confidence interval is below the action level. Otherwise, the opposite conclusion is reached. Example 5.13 demonstrates the application of this test for deciding whether the waste is hazardous or not.

Example 5.13:   One-sample Student's t-test

Excavated soil that is suspected of containing lead has been stockpiled. We may use this soil as backfill (i.e. place it back into the ground), if the stockpile mean lead concentration is below the action level of 100 mg/kg with a 95 percent confidence level. The DQO development for this problem is presented in Examples 2.1, 2.2, and 2.3.

In Example 2.3, we have calculated that 14 samples are needed to reach the decision with a 95 percent level of confidence. To be on the safe side, we collected and analyzed 20 samples. The collected samples have the concentrations of lead ranging from 5 to 210 mg/kg; the mean concentration is 86 mg/kg; the standard deviation is 63 mg/kg; and the standard error is 14 mg/kg. From Appendix 1, Table 2 we determine that the t-value for 19 degrees of freedom (the number of samples less one) and a one-sided confidence interval for $\alpha = 0.05$ is 1.729. Entering these data into Equation 10, Appendix 1, we calculate the 95 percent confidence interval of the mean: 86 ± 24 mg/kg. The upper limit of the confidence interval is 110 mg/kg and it exceeds the action level. Therefore, the null hypothesis $H_0 : \mu \geq 100$ mg/kg, formulated in Example 2.2 is true, as supported by the sample data. Based on this calculation we make a decision not to use the soil as backfill.

What happens to the same data set, if we reduce the confidence level?

- For a 90 percent confidence level, the upper limit of the confidence interval is 104 mg/kg, still exceeding the action level.
- For an 80 percent confidence level, the upper limit of the confidence interval is 98 mg/kg; now it is below the action level.

This example illustrates the profound effect the selection of the confidence level has on project decisions.

All detected values are used in statistical calculations, unqualified and estimated alike. Undetected compounds that may be present in a data set must also be included in a statistical evaluation. If the number of undetected compounds is relatively small (less than 15 percent of the total number of compounds), the RL value or one half of it may be used in calculations (EPA, 1997a). Various methods of statistical analysis may be applied for a greater proportion of undetected compounds.

If a data set is insufficiently complete due to various sampling and non-sampling errors, the test may show that more samples would be needed in order to achieve the project objectives with a stated level of confidence. In this case, additional samples may be collected to fill the data gaps.

5.3.7 Step 7—Prepare Data Quality Assessment Report

Data Quality Assessment Report (DQAR) is the last stone placed at the top of the data collection pyramid. The DQAR summarizes the data collection activities and states whether the data are of the right type, quantity, and quality to support their intended use.

The DQAR may be part of a general report that details all project activities, findings, and conclusions, or be a separate document that becomes part of the permanent project record. Typically, the chemist who conducts DQA prepares the DQAR. The information needed for the DQAR preparation comprises field records, laboratory data, data evaluation summaries, sample data tabulated according to their use, and results of statistical testing.

Depending on the project objectives and contractual conditions, DQARs may be lengthy and elaborate or short and concise. *Important is that a DQAR is not an enumeration of field and laboratory deficiencies and that it formulates the DQA conclusions on the data validity and relevancy in the context of the intended use of the data.*

A DQAR may be prepared according to the following outline:

1. Introduction

    - Short description of the project objectives and of the intended data use
    - References to the SAP and other project documents used in data collection

2. Field Activities

    - Summary of sampling, including a description of the types and numbers of samples and analytical method requirements for each intended use
    - Sample Tracking Log

3. Laboratory Analysis

    - Names and addresses of analytical laboratories
    - Statement of laboratory accreditations
    - References to guidance documents for analytical methods

4. Discussion of Data Quality

- Short description of the evaluation procedure with the references to the source of acceptance criteria
- Explanation of data qualifier use
- Tabulations of qualified data for each intended use, as described in Example 5.7.

5. Accuracy and precision of data for each intended use

- Summary statement on accuracy and precision with exceptions interpreted in context of the contaminant of concern

6. Representativeness for each intended use

- Summary statement on the observance of sample handling requirements (holding time; container; preservation; trip and equipment blanks) with exceptions interpreted in context of the contaminant of concern
- Conclusions of the field duplicate evaluation

7. Comparability

- Qualitative evaluation of comparability

8. Completeness for each intended use

- Results of completeness calculations with the emphasis on completeness of relevant data

9. Statistical evaluation of the data collected for each intended use

- Results of statistical evaluation
- Recommendations based on statistical evaluation

10. Conclusion

- Summary of deviations from the SAP and their effects on the collected data type, quality, and quantity
- Summary statement on data validity and relevancy for each intended data use

The emphasis of a DQAR prepared to this outline is on the DQIs (the PARCC parameters), their acceptance criteria, and their meaning for data usability. While focusing on data validity and relevancy relative to the project DQOs, the DQAR, in fact, establishes whether the data collection process has been successful in each of its three phases—planning, implementation and assessment.

Practical Tips: The seven steps of the data quality assessment

1. To assure the correctness of the SAP implementation, a chemist (often called *sample coordinator*) coordinates sampling crews in the field and keeps track of the collected samples.

2. As an alternative, field crews may communicate with the chemist in the office by telephone and by fax for immediate resolution of questions and problems that may arise during sampling.

3. Obtain the data package clarifications and corrections from the laboratory as soon as possible after the package has been delivered: retrieval of archived laboratory data is a lengthy process that may significantly delay data evaluation.

4. Send a copy of the SAP to the data validation company that will evaluate project data.

5. Remember that acceptance criteria of the National Functional Guidelines for Organic and Inorganic Data review apply *only* for the CLP SOW methods.

6. Verify the reported target analyte lists: analytes erroneously omitted from the list reduce analytical completeness.

7. If high sample variability is evident from field duplicate samples, expect poor data comparability.

8. Archive all analytical data after the project activities have been completed and as the project enters into the closeout phase.

9. Clearly identify data packages; field records; records of communication with the laboratory; data evaluation checklists; and summary reports and archive them together for future reference.

Appendices

Appendix 1
Definitions of Basic Statistical Terms

Statistical evaluations of data are warranted by the fact that the true mean concentration μ (the population mean) will never be known and that we can only estimate it with a sample mean $\bar{x}$. As a reflection of this fact, there are two parallel systems of symbols. The attributes of the theoretical distribution of mean concentrations are called **parameters** (true mean μ, variance σ^2, and standard deviation σ). The estimates of parameters obtained from sample results are called **statistic** (sample mean $\bar{x}$, variance s^2, and standard deviation s).

Variable—x	
Mean of possible measurements of variable (the population mean, true mean)—μ *Individual measurement of variable—x_i*	
1	$\mu = \dfrac{\sum\limits_{i=1}^{N} x_i}{N}$, with N—number of possible measurements
Mean of measurements generated by sample (sample mean)—$\bar{x}$	
2	Sample mean for simple random sampling and systematic random sampling: $\bar{x} = \dfrac{\sum\limits_{i=1}^{n} x_i}{n}$, with n—number of sample measurements Sample mean for stratified random sampling:
3	$\bar{x} = \sum\limits_{k=1}^{r} W_k \bar{x}_k$, with $\bar{x}_k$—stratum mean; W_k—fraction of population represented by stratum k; k—number of strata in the range from 1 to r
Population variance—σ^2 and variance of sample—s^2	
4	Sample variance for simple random sampling and systematic random sampling: $s^2 = \dfrac{\sum\limits_{i=1}^{n} x_i^2 - \dfrac{\left(\sum\limits_{i=1}^{n} x_i\right)^2}{n}}{n-1}$ Sample variance for stratified random sampling:
5	$s^2 = \sum\limits_{k-1}^{r} W_k s_k^2$, with s_k^2—stratum variance; W_k—fraction of population represented by stratum k; k—number of strata in the range from 1 to r

Population standard deviation—σ and standard deviation of sample—s	
6	$s = \sqrt{s^2}$

Standard error of sample (standard error of mean)—$s_{\bar{x}}$	
7	$s_{\bar{x}} = \dfrac{s}{\sqrt{n}}$

Statistical tests of environmental chemical data enable us to compare two sets of data to each other or to compare a set to an action level and make decisions based on these comparisons with a chosen level of confidence. Statistical tests are usually based on the assumptions that the measurements are normally distributed on a Gaussian curve and defined by the standard deviation σ, as shown in Figure 1, and that the errors of measurements are random and independent (each given error affects a measurement but does not affect others).

Confidence level establishes the probability of making a correct decision. For example, a 95 percent confidence level means that a correct decision will be made 95 times out of a hundred and an incorrect decision will be made five times out of a hundred. Confidence level is defined as a function of probability α: if α is 0.05, the confidence level is 95 percent; or if α is 0.20, the confidence level is 80 percent.

Confidence limits are a range of values, within which the true mean concentration is expected to lie. Confidence interval is a *range of values within confidence limits* that is expected to capture the true mean concentration *with a chosen probability*. For example, a 95 percent confidence interval of the mean is a range of sample concentrations that will capture the true mean concentration 95 percent of the time,

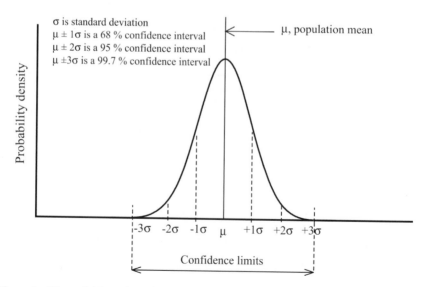

Figure 1   Normal (Gaussian) distribution, confidence limits, and confidence intervals.

but this value will be different every time. The concepts of confidence limits and confidence intervals are illustrated in Figure 1.

Student's t-test is frequently used in statistical evaluations of environmental chemical data. It establishes a relationship between the mean ($\bar{x}$) of normally distributed sample measurements, their sample standard deviation (s), and the population mean (μ). Confidence intervals may be calculated based on Student's t-test (Equation 10). The upper limit of the confidence interval is compared to the action level to determine whether the sampled medium contains a hazardous concentration of a pollutant. If the upper confidence limit is below the action level, the medium is not hazardous; otherwise the opposite conclusion is reached.

Probability, level of significance—a	
Confidence level, percent—(1−a) × 100	
Degrees of freedom—df	
8	$df = n - 1$
Critical value—t	
Student's t-test	
9	$t = \dfrac{\bar{x}-\mu}{s_{\bar{x}}}$
Confidence interval for μ—CI	
10	$CI = \bar{x} \pm t_a s_{\bar{x}}$ with t_a obtained from Table 1 for appropriate degrees of freedom
Action level—C_a	
Appropriate number of samples to collect from a solid waste (financial constraints not considered)—n	
11	$n = \dfrac{t_a^2 s^2}{(C_a - \bar{x})^2}$ with one-sided t_a obtained from Table 1 for appropriate degrees of freedom

Statistical tests may be *one-sided* (*one-tailed*) or *two-sided* (*two-tailed*). One-sided confidence intervals are used for testing the data that are compared to action levels to determine whether the mean concentration is greater or lower than the action level. Two-sided confidence limits are used for comparing two sets of data to each other to establish whether they differ, for example, for comparing sample concentrations to background concentrations. One-sided and two-sided confidence intervals are illustrated in Figure 2.

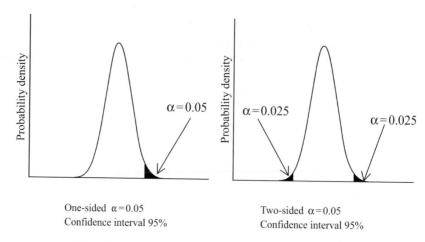

One-sided $\alpha = 0.05$
Confidence interval 95%

Two-sided $\alpha = 0.05$
Confidence interval 95%

Figure 2   One-sided and two-sided confidence intervals.

Table 1 Tabulated values of Student's 't' critical values

Degrees of freedom $df = (n-1)$	α for one-sided confidence interval								
	0.30	0.25	0.20	0.15	0.10	0.05	0.025	0.01	0.005
	One-sided confidence interval								
	70%	75%	80%	85%	90%	95%	97.5%	99%	99.5%
	α for determining two-sided confidence interval								
	0.60	0.50	0.40	0.30	0.20	0.10	0.05	0.02	0.01
	Two-sided confidence interval								
	40%	50%	60%	70%	80%	90%	95%	98%	99%
	t values								
1	0.727	1.000	1.376	1.963	3.078	6.314	12.706	31.821	63.657
2	0.617	0.816	1.061	1.386	1.886	2.920	4.303	6.965	9.925
3	0.584	0.765	0.978	1.250	1.638	2.353	3.182	4.541	5.841
4	0.569	0.741	0.941	1.190	1.533	2.132	2.776	3.747	4.604
5	0.559	0.727	0.920	1.156	1.476	2.015	2.571	3.365	4.032
6	0.553	0.718	0.906	1.134	1.440	1.943	2.447	3.143	3.707
7	0.549	0.711	0.896	1.119	1.415	1.895	2.365	2.998	3.499
8	0.546	0.706	0.889	1.108	1.397	1.860	2.306	2.896	3.355
9	0.543	0.703	0.883	1.100	1.383	1.833	2.262	2.821	3.250
10	0.542	0.700	0.879	1.093	1.372	1.812	2.228	2.764	3.169
11	0.540	0.697	0.876	1.088	1.363	1.796	2.201	2.718	3.106
12	0.539	0.695	0.873	1.083	1.356	1.782	2.179	2.681	3.055
13	0.538	0.694	0.870	1.079	1.350	1.771	2.160	2.650	3.012
14	0.537	0.692	0.868	1.076	1.345	1.761	2.145	2.624	2.977
15	0.536	0.691	0.866	1.074	1.340	1.753	2.131	2.602	2.947
16	0.535	0.690	0.865	1.071	1.337	1.746	2.120	2.583	2.921
17	0.534	0.689	0.863	1.069	1.333	1.740	2.110	2.567	2.898
18	0.534	0.688	0.862	1.067	1.330	1.734	2.101	2.552	2.878
19	0.533	0.688	0.861	1.066	1.328	1.729	2.093	2.539	2.861
20	0.533	0.687	0.860	1.064	1.325	1.725	2.086	2.528	2.845
21	0.532	0.686	0.859	1.063	1.323	1.721	2.080	2.518	2.831
22	0.532	0.686	0.858	1.061	1.321	1.717	2.074	2.508	2.819
23	0.532	0.685	0.858	1.060	1.319	1.714	2.069	2.500	2.807
24	0.531	0.685	0.857	1.059	1.318	1.711	2.064	2.492	2.797
25	0.531	0.684	0.856	1.058	1.316	1.708	2.060	2.485	2.787
26	0.531	0.684	0.856	1.058	1.315	1.706	2.056	2.479	2.779
27	0.531	0.684	0.855	1.057	1.314	1.703	2.052	2.473	2.771
28	0.530	0.683	0.855	1.056	1.313	1.701	2.048	2.467	2.763
29	0.530	0.683	0.854	1.055	1.311	1.699	2.045	2.462	2.756
30	0.530	0.683	0.854	1.055	1.310	1.697	2.042	2.457	2.750
40	0.529	0.681	0.851	1.050	1.303	1.684	2.021	2.423	2.704
60	0.527	0.679	0.848	1.046	1.296	1.671	2.000	2.390	2.660
120	0.526	0.677	0.845	1.041	1.289	1.658	1.980	2.358	2.617
∞	0.524	0.674	0.842	1.036	1.282	1.645	1.960	2.326	2.576

Appendix 2
Toxic and Conventional Pollutants for the Clean Water Act National Pollutant Discharge Elimination System[1]

		Toxic Pollutants		
1	Acenaphthene		34	Endrin and metabolites
2	Acrolein		35	Ethylbenzene
3	Acrylonitrile		36	Fluoranthene
4	Aldrin/Dieldrin		37	Haloethers[5]
5	Antimony and compounds[2]		38	Halomethanes[6]
6	Arsenic and compounds		39	Heptachlor and metabolites
7	Asbestos		40	Hexachlorobutadiene
8	Benzene		41	Hexachlorocyclohexane
9	Benzidine		42	Hexachlorocyclopentadiene
10	Beryllium and compounds		43	Isophorone
11	Cadmium and compounds		44	Lead and compounds
12	Carbon tetrachloride		45	Mercury and compounds
13	Chlordane (technical mixture and metabolites)		46	Naphthalene
14	Chlorinated benzenes (other than dichlorobenzenes)		47	Nickel and compounds
15	Chlorinated ethanes[3]		48	Nitrobenzene
16	Chloroalkyl ethers (chloroethyl and mixed ethers)		49	Nitrophenols (including 2,3-dinitrophenol, dinitrocresol)
17	Chlorinated naphthalene		50	Nitrosamines
18	Chlorinated phenols[4]		51	Pentachlorophenol
19	Chloroform		52	Phenol
20	2-Chlorophenol		53	Phthalate esters
21	Chromium and compounds		54	Polychlorinated biphenyls
22	Copper and compounds		55	Polynuclear aromatic hydrocarbons[7]
23	Cyanides		56	Selenium and compounds
24	DDT and metabolites		57	Silver and compounds
25	Dichlorobenzenes (1,2-, 1,3-, 1,4-dichlorobenzenes)		58	2,3,7,8-Tetrachlorodibenzo-*p*-dioxin
26	Dichlorobenzidine		59	Tetrachloroethylene
27	Dichloroethylenes (1,1- and 1,2-dichloroethylene)		60	Thallium and compounds
28	2,4-Dichlorophenol		61	Toluene
29	Dichloropropane and dichloropropene		62	Toxaphene
30	2,4-Dimethylphenol		63	Trichloroethylene
31	Dinitrotoluene		64	Vinyl chloride
32	Diphenylhydrazine		65	Zinc and compounds
33	Endosulfan and metabolites			

Appendix 2   (Continued)

Conventional Pollutants
1   Biochemical oxygen demand
2   Total suspended solids, nonfilterable
3   pH
4   Fecal coliform
5   Oil and grease

[1] EPA, 1998b.

[2] The term *compound* includes organic and inorganic compounds.

[3] Includes 1,2-dichloroethane, 1,1,1-trichloroethane, and hexachloroethane.

[4] Other than those listed elsewhere; includes trichlorophenols and chlorinated cresols.

[5] Other than those listed elsewhere; includes chlorophenylphenyl ethers, bromophenylphenyl ethers, bis(dichloroisopropyl) ether, bis(chloroethoxy)methane, and polychlorinated diphenyl ethers.

[6] Other than those listed elsewhere; includes methylene chloride, methylchloride, methylbromide, bromoform, dichlorobromomethane.

[7] Include benzanthracenes, benzopyrenes, benzofluoranthene, chrysenes, dibenzanthracenes, and indenopyrenes.

Appendix 3
Maximum Contaminant Concentrations for the Toxicity Characteristic[1]

EPA Hazardous Waste Number	Contaminant	Chemical Abstract Service Number	TCLP Regulatory Level (mg/l)
D004	Arsenic	7440–38–2	5.0
D005	Barium	7440–39–3	100.0
D018	Benzene	71–43–2	0.5
D006	Cadmium	7440–43–9	1.0
D019	Carbon tetrachloride	56–23–5	0.5
D020	Chlordane	57–74–9	0.03
D021	Chlorobenzene	108–90–7	100.0
D022	Chloroform	67–66–3	6.0
D007	Chromium	7440–47–3	5.0
D023	*o*-Cresol	95–48–7	200.0
D024	*m*-Cresol	108–39–4	200.0
D025	*p*-Cresol	106–44–5	200.0
D026	Total cresol		200.0[2]
D016	2,4-Dichlorophenoxy-acetic acid (2,4-D)	94–75–7	10.0
D027	1,4-Dichlorobenzene	106–46–7	7.5
D028	1,2-Dichloroethane	107–06–2	0.5
D029	1,1-Dichloroethylene	75–35–4	0.7
D030	2,4-Dinitrotoluene	121–14–2	0.13
D012	Endrin	72–20–8	0.02
D031	Heptachlor and heptachlor epoxide	76–44–8	0.008
D032	Hexachlorobenzene	118–74–1	0.13
D033	Hexachlorobutadiene	87–68–3	0.5
D034	Hexachloroethane	67–72–1	3.0
D008	Lead	7439–92–1	5.0
D013	Lindane	58–89–9	0.4
D009	Mercury	7439–97–6	0.2
D014	Methoxychlor	72–43–5	10.0
D035	Methyl ethyl ketone	78–93–3	200.0
D036	Nitrobenzene	98–95–3	2.0
D037	Pentachlorophenol	87–86–5	100.0
D038	Pyridine	110–86–1	5.0
D010	Selenium	7782–49–2	1.0
D011	Silver	7440–22–4	5.0
D039	Tetrachloroethylene	127–18–4	0.7
D015	Toxaphene	8001–35–2	0.5
D040	Trichloroethylene	79–01–6	0.5
D041	2,4,5-Trichlorophenol	95–95–4	400.0
D042	2,4,6-Trichlorophenol	88–06–2	2.0
D017	2,4,5-Trichlorophenoxy-propionic acid (2,4,5-TP or Silvex)	93–72–1	1.0
D043	Vinyl chloride	75–01–4	0.2

[1] EPA, 1990.
[2] If *ortho*-, *meta*- and *para*-cresol concentrations cannot be differentiated, the total cresol concentration is used.

Appendix 4
List of California Toxic Substances, Their Soluble Threshold Limit Concentrations and Total Threshold Limit Concentrations[1]

Substance	STLC (mg/l)	TTLC[2] (mg/kg)
Inorganic Persistent and Bioaccumulative Toxic Substances[3]		
Antimony and/or antimony compounds	15	500
Arsenic and/or arsenic compounds	5.0	500
Asbestos		1.0%
Barium and/or barium compounds (excluding barite)	100	10,000
Beryllium and/or beryllium compounds	0.75	75
Cadmium and/or cadmium compounds	1.0	100
Chromium (VI) compounds	5	500
Chromium and/or chromium (III) compounds	5	2,500
Cobalt and/or cobalt compounds	80	8,000
Copper and/or copper compounds	25	2,500
Fluoride salts	180	18,000
Lead and/or lead compounds	5.0	1,000
Mercury and/or mercury compounds	0.2	20
Molybdenum and/or molybdenum compounds	350	3,500
Nickel and/or nickel compounds	20	2,000
Selenium and/or selenium compounds	1.0	100
Silver and/or silver compounds	5	500
Thallium and/or thallium compounds	7.0	700
Vanadium and/or vanadium compounds	24	2,400
Zinc and/or zinc compounds	250	5,000
Organic Persistent and Bioaccumulative Toxic Substances		
Aldrin	0.14	1.4
Chlordane	0.25	2.5
DDT, DDE, DDD	0.1	1.0
2,4-Dichlorophenoxyacetic acid (2,4-D)	10	100
Dieldrin	0.8	8.0
Dioxin (2,3,7,8-TCDD)	0.001	0.01
Endrin	0.02	0.2
Heptachlor	0.47	4.7
Kepone	2.1	21
Organic lead compounds	—	13
Lindane	0.4	4.0
Methoxychlor	10	100
Mirex	2.1	21
Pentachlorophenol	1.7	17
Polychlorinated biphenyls	5.0	50
Toxaphene	0.5	5
Trichloroethylene	204	2,040
2,4,5-Trichlorophenoxypropionic acid (2,4,5-TP or Silvex)	1.0	10

STLC denotes Soluble Threshold Limit Concentration.
TTLC denotes Total Threshold Limit Concentration.
DDD denotes Dichlorodiphenyldichloroethane.
DDE denotes Dichlorodiphenyldichloroethylene.
DDT denotes Dichlorodiphenyltrichloroethane.
TCDD denotes Tetrachlorodibenzodioxin.
[1] CCR, 1991.
[2] Determined on a wet weight basis.
[3] STLC and TTLC values are calculated for the concentrations of elements, not compounds.

Appendix 5
PCB Cleanup Levels for Various Matrices[1]

Cleanup Levels for Bulk PCB Remediation Waste[2]	
High occupancy (e.g. residential) areas	$\leq 1\,\mu g/g$ $> 1\,\mu g/g$ and $\leq 10\,\mu g/g$ if capped
Low occupancy (e.g. electrical substation) areas	$\leq 25\,\mu g/g$ $> 25\,\mu g/g$ and $\leq 50\,\mu g/g$ if fenced $> 25\,\mu g/g$ and $\leq 100\,\mu g/g$ if capped

Cleanup Level for Non-Porous Surfaces in Contact with Liquid PCBs	
High occupancy areas; surfaces destined for reuse	$\leq 10\,\mu g/100\,cm^2$
Low occupancy areas; disposal by smelting	$< 100\,\mu g/100\,cm^2$

Cleanup Level for Non-Porous Surfaces in Contact with Non-Liquid PCBs	
Surfaces destined for reuse	NACE Visual Standard No. 2, Near White Blast Cleaned Surface Finish
Surfaces destined for smelting	NACE Visual Standard No. 3, Commercial Blast Cleaned Surface Finish

Cleanup Levels for Porous Surfaces	
Fresh spill to concrete (within 72 hours of spill)	$\leq 10\,\mu g/100\,cm^2$

Cleanup Level for Liquids	
Water for unrestricted use	$\leq 0.5\,\mu g/l$
Non-contact use in a closed system	$< 200\,\mu g/l$
Water discharged to a treatment works or to navigable waters	$< 3\,\mu g/l$
Organic and non-aqueous inorganic liquids	$\leq 2\,mg/kg$

NACE denotes National Association of Corrosion Engineers.
[1] EPA, 1998c.
[2] Includes soil, sediment, dredged materials, mud, sewage and industrial sludge.

Appendix 6
The EPA Quality Assurance Project Plan Table of Contents

Title Page
Approval Page
Distribution List
List of Tables
List of Figures
A. **Project Management**
    1. Project Task/Organization
    2. Problem Definition/Background
    3. Project/Task description
    4. Data Quality Objectives
        4.1 Project Quality Objectives
        4.2 Measurement Performance Data
    5. Documentation and Records
B. **Measurement Data Acquisition**
    6. Sampling Process Design
    7. Analytical Methods Requirements
        7.1 Organic Compound Analysis
        7.2 Inorganic Compound Analysis
        7.3 Process Control Monitoring
    8. Quality Control Requirements
        8.1 Field QC Requirements
        8.2 Laboratory QC Requirements
    9. Instrument Calibration and Frequency
    10. Data Acquisition Requirements
    11. Data Management
C. **Assessment and Oversight**
    12. Assessment and Response Actions
        12.1 Technical Systems Audits
        12.2 Performance Evaluation Audits
    13. Reports to Management
D. **Data Validation and Usability**
    14. Data Review, Validation, and Verification Requirements
    15. Reconciliation with Data Quality Objectives
        15.1 Assessment of Measurement Performance
        15.2 Data Quality Assessment

Appendix 7
Example of a Sampling and Analysis Plan Table of Contents

Title Page
Approval Page
List of Figures
List of Tables
List of Acronyms and Abbreviations
A.   Project Management
    1.   Introduction
    2.   Site History and Background
    3.   Problem Definition
    4.   Project Organization
    5.   Project Objectives and Intended Use of the Data
B.   Field and Laboratory Data Collection
    6.   Sampling Process Design
    7.   Investigation-Derived Waste
    8.   Field Quality Control
    9.   Sampling Method Requirements
    10.   Analytical Method Requirements
C.   Assessment/Oversight
    11.   Field and Laboratory Audits
D.   Data Validation and Usability
    12.   Data Quality Assessment

Appendix 8
Example of a Field Sampling Supplies and Equipment Checklist

Supplies and Equipment	Quantity	Notes
Sample Collection		
Sample jars, 4-ounce amber glass, wide-mouth, Teflon®-lined lid	❑	
Sample jars, 8-ounce clear glass, wide-mouth, Teflon®-lined lid	❑	
Disposable plastic scoops	❑	
Sample labels	❑	
Clear tape	❑	
Disposable sampling gloves	❑	
Scissors	❑	
Paper towels	❑	
Decontamination Supplies		
Phosphate-free detergent	❑	
Plastic buckets	❑	
Brushes	❑	
Analyte-free water for rinsing	❑	
Documentation		
Blank chain-of-custody forms	❑	
Black ball point pens	❑	
Permanent markers	❑	
Blank field logbook	❑	
Blank sample tracking log	❑	
Photo camera	❑	
Packing and Shipment		
Shipping labels	❑	
Strapping tape	❑	
Ziploc® bags for sample packing	❑	
One-gallon Ziploc® bags for ice	❑	
Packing material	❑	
Shipping labels with laboratory address	❑	
Temperature blanks	❑	

Appendix 9
Example of a Chain-of-Custody Form

Chain-of-Custody Form

Project Number:	Project Name:					Request for Analysis									Chain of Custody Number
Sampler's *(Signature)*															Page _____ of _____
Field Sample ID	Date	Time	Composite	Grab	Sample Location	No. of Containers									Additional Requirements
Relinquished by: *(Signature)*			Date and Time:			Received by: *(Signature)*										Date and Time:
Relinquished by: *(Signature)*			Date and Time:			Received by: *(Signature)*										Date and Time:
Relinquished by: *(Signature)*			Date and Time:			Received by: *(Signature)*										Date and Time:
Comments, special instructions															For Laboratory Use Only

Appendix 10
Example of a Custody Seal and a Sample Label

Custody Seal

Date:_____

Signature:_____

Sample Label

Project name:_____

Date:_____

Time:_____

Sample ID:_____

Sample location: _____

Analysis:_____

Chemical Preservation:_____

Sampler's name: _____

Appendix 11
Example of a Sample Tracking Log

Field Sample ID	Laboratory Sample ID	Date collected	Sample Location	Sampler's Name	Matrix	Analysis	Laboratory Name	Sample Type
S-001	211013-01	10/10/2001	SB-01, 3 ft	J. Smith	Soil	8270, 6010	Laboratory A	Field sample
S-002	211013-02	10/10/2001	SB-01, 3 ft	J. Smith	Soil	8270, 6010	Laboratory A	Field duplicate of S-001
S-003	211013-03	10/10/2001	SB-01, 6 ft	J. Smith	Soil	8270, 6010	Laboratory A	Field sample
S-004	211013-04	10/10/2001	SB-02, 3 ft	J. Smith	Soil	8270, 6010	Laboratory A	Field sample
S-005	211013-05	10/10/2001	SB-02, 6 ft	J. Smith	Soil	8270, 6010	Laboratory A	Field sample
S-006	211013-06	10/10/2001	SB-03, 3 ft	J. Smith	Soil	8270, 6010	Laboratory A	Field sample
S-007	211013-07	10/10/2001	SB-03, 6 ft	J. Smith	Soil	8270, 6010	Laboratory A	Field sample
S-008	211013-08	10/10/2001	SB-04, 3.5 ft	J. Smith	Soil	8270, 6010	Laboratory A	Field sample
S-009	211013-09	10/10/2001	SB-04, 7 ft	J. Smith	Soil	8270, 6010	Laboratory A	Field sample
S-010	211013-10	10/10/2001	SB-05, 3 ft	J. Smith	Soil	8270, 6010	Laboratory A	Field sample
S-011	211013-11	10/10/2001	SB-05, 6 ft	J. Smith	Soil	8270, 6010	Laboratory A	Field sample
S-012	211013-12	10/10/2001	SB-06, 3 ft	J. Smith	Soil	8270, 6010	Laboratory A	Field sample
S-013	211013-13	10/10/2001	SB-06, 3 ft	J. Smith	Soil	8270, 6010	Laboratory A	Field duplicate of S-013
MW-1-014	211056-01	10/12/2001	MW-1	T. Walker	Groundwater	8260	Laboratory B	Field sample
MW-1-015	211056-02	10/12/2001	MW-1	T. Walker	Groundwater	8260	Laboratory B	Field duplicate of MW-1-014
GW-016	211056-03	10/12/2001	NA	T. Walker	Water	8260	Laboratory B	Trip blank
MW-2-017	211056-04	10/12/2001	MW-2	T. Walker	Groundwater	8260	Laboratory B	Field sample
MW-3-018	211056-05	10/12/2001	MW-3	T. Walker	Groundwater	8260	Laboratory B	Field sample
MW-4-019	211056-06	10/12/2001	MW-4	T. Walker	Groundwater	8260	Laboratory B	Field sample
WW-020	211056-07	10/13/2001	Drum No. 1	T. Walker	Purge water	8260, 8270, 6010, 8015	Laboratory B	Field sample

Appendix 12
Analytical Method References, Containers, Preservation, and Holding Time for Soil Samples

Analysis	Extraction and Analysis Methods	Container	Sample size[1]	Preservation	Holding Time
Organic Compounds					
Total petroleum hydrocarbons as gasoline and other volatile petroleum products	SW-846 5035/8015	G, ACD	5 to 25 g	2 to 6°C, NaHSO$_4$ to pH <2 Methanol	48 h for unpreserved samples 7 days for samples preserved, with NaHSO$_4$ 14 days for samples preserved, with methanol
Total petroleum hydrocarbons as diesel fuel and other semivolatile or non-volatile petroleum products	SW-846 3540/8015 SW-846 3550/8015	G, B, S	50 g	2 to 6°C	14 days for extraction, 40 days for analysis
Volatile organic compounds	SW-846 5035/8260 CLP SOW	G, ACD	5 to 25 g	2 to 6°C NaHSO$_4$ to pH <2 Methanol	48 h for unpreserved samples 7 days for samples preserved with NaHSO$_4$ 14 days for samples preserved with methanol
Semivolatile organic compounds	SW-846 3540/8270 SW-846 3550/8270 CLP SOW	G, B, S	50 g	2 to 6°C	14 days for extraction, 40 days for analysis
Halogenated and aromatic volatile compounds	SW-846 5035/8021	G, ACD	5 to 25 g	2 to 6°C NaHSO$_4$ to pH <2 Methanol	48 h for unpreserved samples 7 days for samples preserved with NaHSO$_4$ 14 days for samples preserved with methanol

Appendix 12   (Continued)

Analysis	Extraction and Analysis Methods	Container	Sample size[1]	Preservation	Holding Time
Organic Compounds					
Organochlorine pesticides	SW-846 3540/8081 SW-846 3550/8081 CLP SOW	G, B, S	50 g	2 to 6°C	14 days for extraction, 40 days for analysis
Polychlorinated biphenyls	SW-846 3540/8082 SW-846 3550/8082 CLP SOW	G, B, S	50 g	2 to 6°C	14 days for extraction, 40 days for analysis
Organophosphorus pesticides	SW-846 3540/8141	G, B, S	50 g	2 to 6°C	14 days for extraction, 40 days for analysis
Chlorinated herbicides	SW-846 3540/8151 SW-846 3550/8151	G, B, S	50 g	2 to 6°C	14 days for extraction, 40 days for analysis
Polynuclear aromatic hydrocarbons	SW-846 3540/8310 SW-846 3550/8310	G, B, S	50 g	2 to 6°C	14 days for extraction, 40 days for analysis
Nitroaromatics and nitroamines (trace explosives)	SW-846 3540/8330 SW-846 3550/8330	G, B, S	100 g	2 to 6°C	14 days for extraction, 40 days for analysis
Dioxins/Furans	SW-846 8280 SW-846 8290	G, B, S	100 g	2 to 6°C	14 days for extraction, 40 days for analysis

Appendix 12 (Continued)

Analysis	Extraction and Analysis Methods	Container	Sample size[1]	Preservation	Holding Time
Inorganic Compounds					
Oil and Grease	SM 5520 (B, C, D&E) SW-846 9071	G, B, S	50 g	2 to 6°C	28 days
Total recoverable petroleum hydrocarbons	SM 5520 (F)	G, B, S	50 g	2 to 6°C	14 days for extraction, 40 days for analysis
Organic lead	CA DHS LUFT	G, B, S	50 g	2 to 6°C	14 days
Metals (except mercury)	SW-846 3050 or 3051 SW-846 6010, 6020, 7000 CLP SOW	G, B, S	10 g	2 to 6°C	Six months
Mercury	SW-846 7471 CLP SOW	G, B, S	10 g	2 to 6°C	28 days
Hexavalent chromium	SW-846 7195, 7196, 7197, 7198, 7199	G, B, S	10 g	2 to 6°C	One month for extraction, 4 days for analysis
Total organic carbon	SW-846 9060	G, B, S	10 g	2 to 6°C	28 days
Total organic halides	SW-846 9020	G, B, S	50 g	2 to 6°C	28 days

Appendix 12 (Continued)

Analysis	Extraction and Analysis Methods	Container	Sample size[1]	Preservation	Holding Time
Waste Characterization					
Reactivity—cyanide	SW-846 Chapter 7, Part 7.3	G, B, S	10 g	2 to 6°C	As soon as possible
Reactivity—sulfide	SW-846 Chapter 7, Part 7.3	G, B, S	10 g	2 to 6°C	As soon as possible
Corrosivity	SW-846 Chapter 7, Part 7.2	G, B, S	50 g	2 to 6°C	As soon as possible
Ignitability	SW-846 Chapter 7, Part 7.1	G, B, S	100 g	2 to 6°C	As soon as possible
Toxicity Characteristic Leaching Procedure (TCLP)	SW-846 1311	G, B, S	—	2 to 6°C	Per individual method
Synthetic Precipitation Leaching Procedure (SPLP)	SW-846 1312	G, B, S	—	2 to 6°C	Per individual method
TCLP or SPLP—VOCs	SW-846 1311 or 1312 SW-846 8260	G, B, S	25 g	2 to 6°C	14 days for extraction, 14 days for analysis
TCLP or SPLP—SVOCs	SW-846 1311 or 1312 SW-846 8270	G, B, S	100 g	2 to 6°C	14 days for extraction, 7 days for preparative extraction, 40 days for analysis
TCLP or SPLP—Pesticides	SW-846 1311 or 1312 SW-846 8081	G, B, S	100 g	2 to 6°C	14 days for extraction, 7 days for preparative extraction, 40 days for analysis

Appendix 12 (Continued)

Analysis	Extraction and Analysis Methods	Container	Sample size[1]	Preservation	Holding Time
		Waste Characterization			
TCLP or SPLP—Herbicides	SW-846 1311 or 1312 SW-846 8151	G, B, S	100 g	2 to 6°C	14 days for extraction, 7 days for preparative extraction, 40 days for analysis
TCLP or SPLP—Metals	SW-846 1311 or 1312 SW-846 6010/7472	G, B, S	100 g	2 to 6°C	Six months for extraction, six months for analysis, except for mercury Mercury: 28 days for extraction, 28 days for analysis
California Waste Extraction Test	CCR Title 22, §66261.126	G, B, S	50 g	2 to 6°C	14 days for extraction
Paint filter liquid test	SW-846 9095	G, B, S	100 g	2 to 6°C	Not specified
Hazardous waste aquatic bioassay	CCR Title 22, 66261.24	G, B, S	50 g	2 to 6°C	Not specified

[1]Sample size represents a weight of soil sufficient to perform one analysis only.
Acronyms and abbreviations:
g denotes gram.
G denotes a 4- or 8-ounce glass jar with a PFTE-lined lid.
ACD denotes airtight coring device.
B denotes brass sleeve.
S denotes steel sleeve.
°C denotes degrees Celsius.
NaHSO$_4$ denotes sodium bisulfate.
Analytical method references:
SW-846 denotes US Environmental Protection Agency, *Test Methods for Evaluating Solid Waste, Physical/Chemical Methods, SW-846.*
CLP SOW denotes US Environmental Protection Agency, *Contract Laboratory Program Statement of Work.*
SM denotes American Public Health Association, *Standard Methods for Examination of Water and Wastewater,* 20th Edition, 1998.
CA DHS LUFT denotes the State of California, Department of Health Services, *Leaking Underground Fuel Tank Field Manual,* 1989.
CCR Title 22 denotes the State of California, *California Code of Regulations Title 22,* 1991.

Appendix 13
Analytical Method Requirements, Sample Containers, Preservation, and Holding Time for Water Samples

Analysis	Extraction and Analysis Methods	Container	Sample size[1]	Preservation[2]	Holding Time
Organic Compounds					
Total petroleum hydrocarbons as gasoline and other volatile petroleum products	SW-846 5030/8015	Three glass 40-ml vials, PTFE-lined septum caps	5 to 25 ml	2 to 6°C, no headspace, HCl, pH <2	14 days
Total petroleum hydrocarbons as diesel fuel and other semivolatile or non-volatile petroleum products	SW-846 3510/8015 SW-846 3520/8015	One liter amber glass, PTFE-lined lid	1000 ml	2 to 6°C	7 days for extraction, 40 days for analysis
Volatile organic compounds	SW-846 5030/8260 CLP SOW	Three glass 40-ml vials, PTFE-lined septum caps	5 to 25 ml	2 to 6°C, HCl pH <2	14 days
Semivolatile organic compounds	SW-846 3510/8270 SW-846 3520/8270 CLP SOW	One liter amber glass, PTFE-lined lid	1000 ml	2 to 6°C	7 days for extraction, 40 days for analysis
Halogenated and aromatic volatile compounds	SW-846 5030/8021	Three glass 40-ml vials, PTFE-lined septum caps	5 to 25 ml	2 to 6°C, HCl pH <2	14 days
Organochlorine pesticides	SW-846 3510/8081 SW-846 3520/8081 CLP SOW	One liter amber glass, PTFE-lined lid	1000 ml	2 to 6°C	7 days for extraction, 40 days for analysis
Polychlorinated biphenyls	SW-846 3510/8082 SW-846 3520/8082 CLP SOW	One liter amber glass, PTFE-lined lid	1000 ml	2 to 6°C	7 days for extraction, 40 days for analysis

Appendix 13 (Continued)

Analysis	Extraction and Analysis Methods	Container	Sample size[1]	Preservation[2]	Holding Time
Organic Compounds					
Organophosphorus pesticides	SW-846 3510/8141 SW-846 3520/8141	One liter amber glass, PTFE-lined lid	1000 ml	2 to 6°C	7 days for extraction, 40 days for analysis
Chlorinated herbicides	SW-846 3510/8151 SW-846 3520/8151	One liter amber glass, PTFE-lined lid	1000 ml	2 to 6°C	7 days for extraction, 40 days for analysis
Polynuclear aromatic hydrocarbons	SW-846 3510/8310 SW-846 3520/8310	One liter amber glass, PTFE-lined lid	1000 ml	2 to 6°C	7 days for extraction, 40 days for analysis
Nitroaromatics and nitroamines (trace explosives)	SW-846 3510/8330 SW-846 3520/8330	One liter amber glass, PTFE-lined lid	1000 ml	2 to 6°C	7 days for extraction, 40 days for analysis
Dioxins/Furans	SW-846 8280 SW-846 8290	One liter amber glass, PTFE-lined lid	1000 ml	2 to 6°C	7 days for extraction, 40 days for analysis
Inorganic Compounds					
Oil and Grease	EPA 1664 SM 5520 (B, C, D&E) EPA 413.1, 413.2	One liter amber glass, PTFE-lined lid	1000 ml	2 to 6°C, H_2SO_4 or HCl pH <2	28 days
Total recoverable petroleum hydrocarbons	EPA 1664 EPA 418.1 SM 5520 (F)	One liter amber glass, PTFE-lined lid	1000 ml	2 to 6°C, HCl pH <2	28 days for extraction, 40 days for analysis
Organic lead	CA DHS LUFT	Four glass 40-ml vials, PTFE-lined septum caps	100 ml	2 to 6°C, no headspace	14 days

Appendix 13 (Continued)

Analysis	Extraction and Analysis Methods	Container	Sample size[1]	Preservation[2]	Holding Time
Inorganic Compounds					
Total metals (except mercury) Dissolved metals: filter in the field through a 0.45 μm filter before adding acid	SW-846 3005 or 3010 SW-846 6010 6020, 7000 EPA 200.0, 200.7 CLP SOW	500 ml HDPE or glass	100 ml	Room temperature HNO$_3$, pH < 2	Six months
Total mercury Dissolved mercury: filter in the field through a 0.45 μm filter before adding acid	SW-846 7472 EPA 245.1, 245.2, 1631 CLP SOW	500 ml HDPE or glass	100 ml	2 to 6°C, HNO$_3$ to pH < 2	28 days
Hexavalent chromium	SW-846 7195, 7196, 7197, 7198, 7199 EPA 218.5	500 ml HDPE or glass	200 ml	2 to 6°C	24 h
Total organic carbon	SW-846 9060 EPA 415.1	500 ml HDPE or glass	25 ml	2 to 6°C, H$_2$SO$_4$ or HCl, pH < 2	28 days
Total organic halides	SW-846 9020	Glass 40-ml vial	50 ml	2 to 6°C, H$_2$SO$_4$ or HCl, pH < 2	28 days
Anions	EPA 300.0	100 ml HDPE	20 ml	2 to 6°C	48 h
Nitrite	EPA 354.1	500 ml HDPE or glass	100 ml	2 to 6°C	48 h
Nitrate	EPA 300.0/353.2	500 ml HDPE or glass	50 ml	2 to 6°C	48 h

[1] Sample size represents a volume of water sufficient to perform one analysis only.

[2] Add 0.008% of sodium thiosulfate ($Na_2S_2O_3$) to all samples collected for organic compound analysis with the exception of SW-846 Methods 8081, 8082, and 8141, if residual chlorine is present.

[3] If residual chlorine or another oxidizer is present, add 0.6 g ascorbic acid per one liter of the sample.

Acronyms and abbreviations:

ml denotes milliliter.
°C denotes degrees Celsius.
HNO_3 denotes nitric acid.
NaOH denotes sodium hydroxide.
HCl denotes hydrochloric acid.
HDPE denotes high-density polyethylene.
H_2SO_4 denotes sulfuric acid.
PTFE denotes polytetrafluoroethylene.

Analytical method references:

SW-846 denotes US Environmental Protection Agency, *Test Methods for Evaluating Solid Waste, Physical/Chemical Methods, SW-846.*
CLP SOW denotes US Environmental Protection Agency, *Contract Laboratory Program Statement of Work for Organic Analysis.*
EPA denotes US Environmental Protection Agency, *Methods for Chemical Analysis of Water and Wastes*, EPA-600/4-79-020, 1983.
SM denotes American Public Health Association, *Standard Methods for Examination of Water and Wastewater*, 20th Edition, 1998.
CA DHS LUFT denotes the State of California, Department of Health Services, *Leaking Underground Fuel Tank Field Manual*, 1989.
Hach denotes Hach Company, Loveland, Colorado.
CCR Title 22 denotes the State of California, *California Code of Regulations Title 22*, 1991.
California Department of Fish and Game denotes the State of California Department of Fish and Game, *Static Acute Bioassay Procedures for Hazardous Waste Samples*, 1988.
EPA-600/4-90/027F denotes US Environmental Protection Agency, *Methods for Measuring the Acute Toxicity of Effluent and Receiving Waters to Freshwater and Marine Organisms*, Fourth Edition, 1993.
EPA-600/4-91/002 denotes US Environmental Protection Agency, *Short-Term Methods for Estimating the Chronic Toxicity of Effluent and Receiving Waters to Freshwater Organisms*, Third Edition, 1994.
EPA-600/4-91/003 denotes US Environmental Protection Agency, *Short-Term Methods for Estimating the Chronic Toxicity of Effluent and Receiving Waters to Estuarine and Marine Organisms*, 2nd Edition, 1994.
ASTM denotes American Society for Testing and Materials, *Annual Book of Standards*, 2000.

Appendix 14
Stockpile Statistics Worksheet

		Stockpile Statistics Worksheet	
Step 1	Number of collected samples	n	$n =$
Step 2	Calculate sample concentration mean $\bar{x}$ $n =$ number of samples	$\bar{x} = \dfrac{\sum\limits_{i=1}^{n} x_i}{n}$	$\bar{x} =$
Step 3	Calculate sample variance	$s^2 = \dfrac{\sum\limits_{i=1}^{n} x_i^2 - \dfrac{\left(\sum\limits_{i=1}^{n} x_i\right)^2}{n}}{n-1}$	$s^2 =$
Step 4	Calculate sample standard deviation	$s = \sqrt{s^2}$	$s =$
Step 5	Calculate degree of freedom	$df = n - 1$	$df =$
Step 6	Calculate standard error of the mean	$s_{\bar{x}} = \dfrac{s}{\sqrt{n}}$	$s_{\bar{x}} =$
Step 7	Obtain *Student's t value* corresponding to the degree of freedom value determined in Step 5	Look up Table 1, Appendix 1, for the selected *t*-value	$t_\alpha =$
Step 8	Calculate the confidence interval	$CI = \bar{x} \pm t_\alpha s_x$	$CI =$
Step 9	Obtain action level (regulatory threshold) C_a for the contaminant of concern	C_a	$C_a =$
Step 10	Calculate the difference	$\Delta = C_a - \bar{x}$	$\Delta =$
Step 11	Determine minimum sufficient number of samples	$n_{min} = \dfrac{t_\alpha^2 s^2}{\Delta^2}$	$n_{min} =$
Step 12	Decide whether the number of samples is sufficient	$n_{min} \geqslant n$	Yes
		$n_{min} < n$	No

The contaminant of concern is not present in the stockpile above the action level if the upper limit of the *CI* is less than the C_a.

Appendix 15
Example of a Sample Container Tracking Log

Sample Container Tracking Log

Project Name and Number:

Balance Calibration Verification (10 g): Containers prepared by:

Balance Calibration Verification (100 g): Sampled by:

Date Sampled	Container No.	Field Sample ID	Container with Methanol	Container with Methanol (Prior to Sampling)	Container with Methanol and Soil	Comments
			Container Weight, grams			

Appendix 16
Example of a Groundwater Sampling Form

<table>
<tr><td colspan="7" align="center">**Groundwater Sampling Form**</td></tr>
<tr><td colspan="7" align="center">General Information</td></tr>
<tr><td colspan="3">Site Location:</td><td colspan="4">Date:</td></tr>
<tr><td colspan="3">Project Name:</td><td colspan="4">Well ID:</td></tr>
<tr><td colspan="3">Project Number:</td><td colspan="4">Well condition:</td></tr>
<tr><td colspan="3">Field Sampler's Name:</td><td colspan="4">PID reading:</td></tr>
<tr><td colspan="3">Weather:</td><td colspan="4">Information recorded by:</td></tr>
<tr><td colspan="7" align="center">Well Detail and Volume Calculation</td></tr>
<tr><td>Well type (monitor, extraction, other)</td><td></td><td>Casing diameter, in</td><td></td><td></td><td rowspan="5">Total purge volume, gallons</td><td rowspan="5">Actual volume purged, gallons</td></tr>
<tr><td>Screened interval depth, ft</td><td></td><td>Well casing diameter, in</td><td></td><td></td></tr>
<tr><td>Total well depth (T), ft</td><td></td><td>F factor, gallons/ft</td><td></td><td></td></tr>
<tr><td>Depth to water (W), ft</td><td></td><td>Number of volumes to purge, N</td><td></td><td></td></tr>
<tr><td>Water column depth (H = T − W) ft</td><td></td><td>Purge volume calculation N×(V=H × F), gallons</td><td></td><td></td></tr>
<tr><td colspan="7" align="center">Purging Information</td></tr>
<tr><td colspan="2">Purge method:</td><td>Bailing ☐</td><td colspan="2">High speed pump ☐</td><td colspan="2">Low-flow micropurge ☐</td></tr>
<tr><td colspan="5">Pump information (type and model):</td><td colspan="2">Pump intake depth, ft</td></tr>
<tr><td colspan="2">Purging pump flow rate, l/min</td><td colspan="5"></td></tr>
<tr><td colspan="2">Purge water storage/disposal</td><td colspan="5"></td></tr>
<tr><td colspan="7" align="center">Groundwater Parameter Measurement</td></tr>
<tr><td colspan="7">Meter model and type:</td></tr>
<tr><td>Time, min</td><td>Purged volume, gallons</td><td>pH</td><td>Temp, °C</td><td>Conductivity, (μS/cm)</td><td>ORP, mV</td><td>DO, mg/liter</td><td>Turbidity, NTU</td></tr>
<tr><td colspan="2">Stabilization criteria:</td><td>±0.1</td><td>±1°C</td><td>±10%</td><td>±1 mV</td><td>±10%</td><td>±10%</td></tr>
<tr><td></td><td></td><td></td><td></td><td></td><td></td><td></td><td></td></tr>
<tr><td></td><td></td><td></td><td></td><td></td><td></td><td></td><td></td></tr>
<tr><td></td><td></td><td></td><td></td><td></td><td></td><td></td><td></td></tr>
<tr><td></td><td></td><td></td><td></td><td></td><td></td><td></td><td></td></tr>
<tr><td></td><td></td><td></td><td></td><td></td><td></td><td></td><td></td></tr>
<tr><td colspan="8">Comments:</td></tr>
<tr><td colspan="8" align="center">Well Sampling</td></tr>
<tr><td colspan="2">Sampling method:</td><td colspan="2">Bailing ☐</td><td colspan="4">Low-flow pump ☐</td></tr>
<tr><td colspan="2">Time of sampling:</td><td colspan="2">Bailer type:</td><td colspan="4">Sampling pump flow rate, ml/min:</td></tr>
<tr><td colspan="2">Sample ID:</td><td colspan="2">Field sample ☐</td><td colspan="2">Field duplicate ☐</td><td colspan="2">Equipment blank ☐</td></tr>
<tr><td>Container/ volume</td><td>Analysis</td><td colspan="2">Preservation</td><td colspan="2">Laboratory</td><td colspan="2">Comments</td></tr>
<tr><td></td><td></td><td colspan="2"></td><td colspan="2"></td><td colspan="2"></td></tr>
<tr><td></td><td></td><td colspan="2"></td><td colspan="2"></td><td colspan="2"></td></tr>
<tr><td></td><td></td><td colspan="2"></td><td colspan="2"></td><td colspan="2"></td></tr>
<tr><td colspan="8">Sampler's signature:</td></tr>
</table>

Appendix 21
Inorganic Data Package Content

Data Package Elements	Validation Level 4	Validation Level 3	Standard Laboratory Report
Case Narrative	✓	✓	
Cross-reference of field sample numbers, laboratory IDs, and laboratory QC batches	✓	✓	
Chain-of-Custody Form, Cooler Receipt Form	✓	✓	✓
Sample and method blank results	✓	✓	✓
LCS/LCSD report (concentration spiked, percent recovered, recovery control limits, the RPD, and RPD control limits)	✓	✓	✓
MS/MSD report (concentration spiked, percent recovered, recovery control limits, the RPD, and RPD control limits)	✓	✓	✓
Laboratory duplicate report and acceptance limits	✓	✓	✓
Initial calibration and acceptance limits	✓	✓ (Summary only)	
Initial and continuing calibration verifications and acceptance limits	✓	✓ (Summary only)	
Sample preparation logs	✓		
Analysis run logs	✓	✓	
Raw data, instrument printouts, copies of bench sheets	✓		
Percent moisture	✓	✓	

LCS/LCSD denotes laboratory control sample/laboratory control sample duplicate.
RPD denotes relative percent difference.
MS/MSD denotes matrix spike/matrix spike duplicate.

Appendix 22
Calculations Used for Compound Quantitation

General Concepts		
1		***Response or Calibration Factor of Linear Calibration*** $$RF_i = \frac{C_{std}}{A_{std}} \text{ or } CF_i = \frac{A_{std}}{C_{std}}$$ RF_i is response factor for an individual calibration standard compound CF_i is calibration factor for an individual calibration standard compound C_{std} is concentration or mass of the calibration standard compound A_{std} is peak area of the calibration standard compound
2		***Average Response or Calibration Factor of Linear Calibration*** $$RF_{average} = \frac{\sum_i^n RF_i}{n} \text{ or } CF_{average} = \frac{\sum_i^n CF_i}{n}$$ $RF_{average}$ or $CF_{average}$ is the average RF or CF of a multipoint calibration n is the number of calibration points
3		***Relative Standard Deviation of Linear Calibration*** $$RSD = \frac{S_{RF}}{RF_{average}} \times 100 \text{ or } RSD = \frac{S_{CF}}{CF_{average}} \times 100$$ RSD is relative standard deviation S_{RF} or S_{CF} is standard deviation of the RFs or CFs of a multipoint calibration
4		***Percent Difference of Linear Calibration*** $$\%D = \frac{RF_{ICV} - RF_{average}}{RF_{average}} \times 100 \text{ or } \%D = \frac{CF_{ICV} - CF_{average}}{CF_{average}} \times 100$$ $\%D$ is percent difference RF_{ICV} or CF_{ICV} is the RF or CF calculated from the ICV analysis
5		***Percent Drift of Non-Linear Calibration*** $$Percent\ Drift = \frac{C_{ICV}^{measured} - C_{ICV}^{true}}{C_{ICV}^{true}} \times 100$$ $C_{ICV}^{measured}$ is measured concentration of the ICV standard C_{ICV}^{true} is true concentration of the ICV standard

Appendix 22   (Continued)

External Standard Calibration Calculations

	Sample Concentration (Direct Injection)
6	$$C_{sample} = \frac{A_{sample} \times RF_{average} \times V_{extract}}{W_{sample}} \times DF \text{ or } C_{sample} = \frac{A_{sample} \times V_{extract}}{CF_{average} \times W_{sample}} \times DF$$ C_{sample} is compound concentration in the sample (μg/kg or μg/l) A_{sample} is peak area of a compound in the sample $RF_{average}$ is the average multipoint calibration RF (μg/ml) $CF_{average}$ is the average multipoint calibration CF (ml/μg) $V_{extract}$ is final extract volume (ml) W_{sample} is weight (kg) or volume (liter) of extracted sample DF is dilution factor
	Sample Concentration (Purge and Trap Water and Low Level Soil)
7	$$C_{sample} = \frac{A_{sample} \times RF_{average}}{W_{sample}} \text{ or } C_{sample} = \frac{A_{sample}}{CF_{average} \times W_{sample}}$$ C_{sample} is compound concentration in the sample (μg/kg or μg/l) A_{sample} is peak area of a compound in the sample $RF_{average}$ or $CF_{average}$ is the average multipoint calibration RF (μg) or CF (μg^{-1}) W_{sample} is weight (kg) or the volume (liter) of the analyzed sample
	Sample Concentration (Purge and Trap Medium and High Level Soil)
8	$$C_{sample} = \frac{A_{sample} \times RF_{average} \times V_{extract}}{W_{sample} \times V_{purged}} \text{ or } C_{sample} = \frac{A_{sample} \times V_{extract}}{CF_{average} \times W_{sample} \times V_{purged}}$$ C_{sample} is compound concentration in the sample (μg/kg) $RF_{average}$ or $CF_{average}$ is the average multipoint calibration RF (μg) or CF (μg^{-1}) A_{sample} is peak area of a compound in the sample $V_{extract}$ is the volume of solvent used for extraction (ml) V_{purged} is the volume of extract purged (ml) W_{sample} is weight of the extracted sample (kg)

Internal Standard Calibration Calculations

	Response Factor of Linear Calibration
9	$$RF_i = \frac{A_{std} \times C_{is}}{A_{is} \times C_{std}}$$ RF_i is response factor for an individual calibration standard compound A_{std} is peak area of the calibration standard compound C_{is} is concentration of the internal standard (μg/ml) A_{is} is peak area of the internal standard C_{std} is concentration of the calibration standard compound (μg/ml)

Appendix 22   (Continued)

	Sample Concentration (Direct Injection)
10	$$C_{sample} = \frac{A_{sample} \times C_{is} \times V_{extract}}{A_{is} \times RF_{average} \times W_{sample}} \times DF$$ C_{sample} is compound concentration in the sample (μg/kg or μg/l) A_{sample} is peak area of a compound in the sample C_{is} is concentration of the internal standard (μg/ml) $V_{extract}$ is final extract volume (ml) A_{is} is peak area or height of the internal standard $RF_{average}$ is the average multipoint calibration RF (unitless) W_{sample} is weight (kg) or volume (liter) of the extracted sample DF is dilution factor
	Sample Concentration (Purge and Trap Water and Low Level Soil)
11	$$C_{sample} = \frac{A_{sample} \times C_{is}}{A_{is} \times RF_{average} \times W_{sample}}$$ C_{sample} is compound concentration in the sample (μg/kg or μg/l) A_{sample} is peak area of a compound in the sample C_{is} is quantity of the internal standard (μg) $RF_{average}$ is the average multipoint calibration RF (unitless) W_{sample} is weight (kg) or volume (liter) of extracted sample
	Sample Concentration (Purge and Trap Medium and High Level Soil)
12	$$C_{sample} = \frac{A_{sample} \times C_{is} \times V_{extract}}{A_{is} \times RF_{average} \times W_{sample} \times V_{purged}}$$ C_{sample} is compound concentration in the sample (μg/kg) $RF_{average}$ is the average multipoint calibration RF (unitless) C_{is} is quantity of the internal standard (μg) A_{sample} is peak area of a compound in the sample $V_{extract}$ is the volume of solvent used for extraction (ml) V_{purged} is the volume of extract purged (ml) W_{sample} is weight of the extracted sample (kg)
	Linear Regression and Non-Linear Calibrations
13	$$C_{sample} = \frac{C_{extract} \times V_{extract}}{W_{sample}}$$ C_{sample} is compound concentration in the sample (μg/kg or μ/l) $C_{extract}$ is compound concentration in the final extract (μg/ml) $V_{extract}$ is total volume of extract (ml) W_{sample} is the weight (kg) or volume (liter) of the extracted or purged sample

Appendix 23 Data Evaluation Checklist—Organic Compound Analysis

Laboratory Project No.: _____ Method: _____ Reviewer: _____ Date: _____

	Yes	No (Describe the cause and list affected samples)	NA
1. Sample Management			
COC dated and signed by all parties			
Holding time and preservation requirements observed			
2. Calibration			
Preparation/analysis methods and analytes correct			
Reporting limits acceptable			
MS tuning acceptable (GC/MS methods only)			
Initial calibration acceptable			
Second source calibration verification acceptable			
CCV(s) acceptable			
Closing CCV acceptable			
Frequency of CCV acceptable			
DDT and Endrin degradation acceptable			
Second column confirmation performed			
3. QC checks			
QC batch cross-reference acceptable			
Method blank acceptable			
Internal standard areas within acceptance limits			
Surrogate standard recoveries within control limits			
LCS within control limits			
LCSD and the RPD within control limits			
MS within control limits			
MSD and the RPD within control limits			
QC checks frequency acceptable			
4. Corrective action required			
If yes, specify:			

NA denotes Not Applicable.

Appendix 24   Data Evaluation Checklist—Trace Element Analysis

Laboratory Project No.:_____   Method:_____   Reviewer:_____   Date:_____

	Yes	No (Describe the cause and list affected samples)	NA
1. Sample Management			
COC dated and signed by all parties			
Holding time and preservation requirements observed			
2. Calibration			
Preparation/analysis methods and analytes correct			
Reporting limits acceptable			
Initial calibration acceptable			
Second source calibration verification acceptable			
CCV(s) acceptable			
Closing CCV acceptable			
Frequency of CCV acceptable			
3. QC checks			
QC batch cross-reference acceptable			
Method blank acceptable			
Post-digestion spike recovery acceptable			
ICP interference check sample acceptable			
Method of standard additions conducted			
Serial dilutions acceptable			
Laboratory duplicate within control limits			
LCS within control limits			
LCSD and the RPD within control limits			
MS within control limits			
MSD and the RPD within control limits			
QC checks frequency acceptable			
4. Corrective action required			
If yes, specify:			

NA denotes Not Applicable.

Appendix 25   Data Evaluation Checklist—Inorganic Compound Analysis

Laboratory Project No.: _____   Method: _____   Reviewer: _____   Date: _____

	Yes	No (Describe the cause and list affected samples)	NA
1. Sample Management			
COC dated and signed by all parties			
Holding time and preservation requirements observed			
2. Calibration			
Preparation/analysis methods and analytes correct			
Reporting limits acceptable			
Initial calibration acceptable			
Second source calibration verification acceptable			
CCV(s) acceptable			
Closing CCV acceptable			
Frequency of CCV acceptable			
3. QC checks			
QC batch cross-reference acceptable			
Method blank acceptable			
Laboratory duplicate within control limits			
LCS within control limits			
LCSD and the RPD within control limits			
MS within control limits			
MSD and the RPD within control limits			
QC checks frequency acceptable			
4. Corrective action required			
If yes, specify:			

NA denotes Not Applicable.

REFERENCES

1. Air Force Center for Environmental Excellence, *Quality Assurance Project Plan Version 3.1*, 2001.
2. Alaska Department of Environmental Conservation (ADEC), *Underground Storage Tanks Procedures Manual*, State of Alaska, 1999.
3. American Public Health Association, *Standard Methods for Examination of Water and Wastewater*, 20th Edition, 1998.
4. American Society for Testing and Materials, *Standard Practice for Estimation of Holding Time for Water Samples Containing Organic Constituents*, D 4515–85, [Annual Book of ASTM Standards, 1987].
5. American Society for Testing and Materials, *Standard Guide for Sampling Waste and Soils for Volatile Organic Compounds*, D 4547–98, [Annual Book of ASTM Standards, 1998].
6. American Petroleum Institute, *Interlaboratory Study of Three Methods for Analyzing Petroleum Hydrocarbons in Soils*, [API Publication Number 4599, March 1994].
7. R. Craven, *A Study of the Effects of Air Shipment on the Volatile Content of Water Samples*, LabLink, Volume 4, Issue 2, [Columbia Analytical Services, 1998].
8. J. Drever, *The Geochemistry of Natural Waters Surface and Groundwater Environments*, 3rd edition, [Prentice Hall, 1997].
9. A.D. Hewitt, *Enhanced Preservation of Volatile Organic Compounds in Soil with Sodium Bisulfate*, USA Cold Regions Research and Engineering Laboratory, Special Report 95–26, [US Army Corps of Engineers, 1995a].
10. A.D. Hewitt, *Chemical Preservation of Volatile Organic Compounds in Soil Subsamples*, USA Cold Regions Research and Engineering Laboratory, Special Report 95–5, [US Army Corps of Engineers, 1995b].
11. A.D. Hewitt, *Storage and Preservation of Soil Samples for Volatile Compound Analysis*, USA Cold Regions Research and Engineering Laboratory, Special Report 99–5, [US Army Corps of Engineers, 1999a].
12. A.D. Hewitt, *Frozen Storage of Soil Samples for VOC Analysis*, Environmental Testing and Analysis, Volume 8, Number 5, September/October 1999b.
13. A.D. Hewitt and N.J.E. Lukash, *Obtaining and Transferring Soils for In-Vial Analysis of Volatile Organic Compounds*, USA Cold Regions Research and Engineering Laboratory, Special Report 96–5, [US Army Corps of Engineers, 1996].
14. International Standardization Organization, *Guide 25, General Requirements for the Competence of Testing and Calibration Laboratories*, 1990.
15. International Standardization Organization, *Standard 17025, General Requirements for the Competence of Testing and Calibration Laboratories*, 1999
16. B. Lesnik and D. Crumbling, *Guidelines for Preparing SAPs Using Systematic Planning and PBMS*, Environmental Testing and Analysis, Volume 10, Number 1, January/February 2001.
17. D.L. Lewis, *Assessing and Controlling Sample Contamination*, in Principles of Environmental Sampling, [edited by Lawrence H. Keith, American Chemical Society, 1988].
18. R. Kipling, *Just So Stories*, [Penguin USA, May 1990].
19. G.H. Kassakhian and S.J. Pacheco, *Alternatives to the Use of ASTM Type II Reagent Water*, Proceedings of the 10th Annual Waste Testing and Quality Assurance Symposium, [US Environmental Protection Agency, 1994].
20. Naval Facilities Engineering Service Center, *Navy Installation Restoration Chemical Data Quality Manual*, 1999.

21. L. Parker, *A Literature Review on Decontaminating Groundwater Sampling Devices, Organic Pollutants*, Special Report 95–14, [Cold Regions Research and Engineering Laboratory, US Army Corps of Engineers, 1995].

22. L. Parker and T. Ranney, *Decontaminating Groundwater Sampling Devices*, Special Report 97–25, Cold Regions Research and Engineering Laboratory, [US Army Corps of Engineers, 1997].

23. E. Popek and G. Kassakhian, *Investigation vs. Remediation: Perception and Reality*, Remediation, Volume 9, Number 1, 1998a.

24. E. Popek and G. Kassakhian, *Declining Chemical Data Quality: A Call to Action*, Proceedings of the 17th Annual National Conference on Managing Quality Systems for Environmental Programs, [US Environmental Protection Agency, 1998b].

25. E. Popek and G. Kassakhian, *The Use and Abuse of QA/QC Samples in Environmental Data Acquisition*, Proceedings of the 18th National Conference on Managing Quality Systems for Environmental Programs, [US Environmental Protection Agency, 1999].

26. R. Puls and M. Barcelona, *Ground Water Sampling for Metals Analysis*, Superfund Ground Water Issue, EPA/540/4–89/001, [US Environmental Protection Agency, 1989].

27. R. Puls and M. Barcelona, *Low-Flow (Minimal Drawdown) Groundwater Sampling Procedure*, EPA/540/S-95/504, [US Environmental Protection Agency, 1995].

28. Orion, *Laboratory Products and Electrochemistry Handbook*, 1999.

29. The State of California, *California Code of Regulations Title 22, Division 4.5, Chapter 11, Article 3, Section 66261.24*, [Barclays California Code of Regulations, Register 91, No. 40, October 4, 1991].

30. The State of Hawai'i, *Technical Guidance Manual for Underground Storage Tank Closure and Release Response*, [State of Hawai'i Department of Health, 1992].

31. D. Turiff and C. Reitmeyer, *Validation of Holding Times for the EnCore™ Sampler*, [En Chem, Inc., 1998].

32. US Army Corps of Engineers, *Requirements for Preparation of Sampling and Analysis Plans*, EM 200–1–3, [US Army Corps of Engineers, 1994].

33. US Department of Defense, *Department of Defense Quality Systems Manual for Environmental Laboratories*, [US Department of Defense, 2000].

34. US Department of Energy Hazardous Waste Remedial Actions Program, *Standard Operating Procedures for Site Characterizations*, [Lockheed Martin Energy Systems, Inc., 1996].

35. US Environmental Protection Agency, *Prescribed Procedures for Measurement of Radioactivity in Water*, EPA 600/4–80–032, [US Environmental Protection Agency, 1980].

36. US Environmental Protection Agency, *Methods for Chemical Analysis of Water and Wastes*, EPA-600/4–79–020, [US Environmental Protection Agency, 1983].

37. US Environmental Protection Agency, *Definition and Procedure for the Determination of the Method Detection Limit*, Code of Federal Regulations Title 40, Part 136, Appendix B, Federal Register Vol. 49, No. 209, [US Government Printing Office, 1984a].

38. US Environmental Protection Agency, *Guidelines for Establishing Test Procedures for the Analysis of Pollutants under the Clean Water Act*, Code of Federal Regulations Title 40, Part 136 Federal Register Vol. 49, No. 209, [US Government Printing Office, 1984b].

39. US Environmental Protection Agency, *National Primary Drinking Water Standard: Synthetic Organics, Inorganics and Bacteriologicals*, Code of Federal Regulations Title 40 Part 141, Federal Register 50, No. 219, [US Government Printing Office, November 18, 1985].

40. US Environmental Protection Agency, *Development of Data Quality Objectives, Description of Stages I and II*, July, [US Environmental Protection Agency, 1986].

41. US Environmental Protection Agency, *Data Quality Objectives for Remedial Response Activities*, EPA/540/G-87/003, [US Environmental Protection Agency, 1987].

42. US Environmental Protection Agency, *Methods for Evaluating the Attainment of Cleanup Standards, Volume 1: Soil and Solid Media*, EPA Report 230/02–89–042, [Office of Policy, Planning and Evaluation, Washington DC, 1989a].

Index